The End of Evolution

The End of Evolution

A Journey in Search of Clues to the Third Mass Extinction Facing Planet Earth

Peter Ward

BANTAM BOOKS
NEW YORK TORONTO LONDON SYDNEY AUCKLAND

THE END OF EVOLUTION
PUBLISHING HISTORY
A Bantam Book
Bantam hardcover edition/July 1994
Bantam trade paperback edition/September 1995

Grateful acknowledgment is made for permission to reprint the following illustrations:

On page 9, from *Earth and Life Through Time,* 2d ed.; on page 36, from *The Proceedings of the American Philosophical Society;* on page 46, by Gregory S. Paul; on page 48, by Gregory S. Paul; on page 83, from *American Scientist.* All of the above originally appeared in *Earth and Life Through Time* by Steven M. Stanley.

Book design by Charles B. Hames
Library of Congress Catalog Card Number: 93-31802.

ISBN 0-553-37469-9

Published simultaneously in the United States and Canada

Bantam Books are published by Bantam Books, a division of Bantam Doubleday Dell Publishing Group, Inc. Its trademark, consisting of the words "Bantam Books" and the portrayal of a rooster, is Registered in U.S. Patent and Trademark Office and in other countries. Marca Registrada. Bantam Books, 1540 Broadway, New York, New York 10036.

PRINTED IN THE UNITED STATES OF AMERICA

BVG 10 9 8 7 6 5 4 3 2 1

For an extraordinary family: Joe and Ruth; Steve, Ann, Nate, and Glynn; Betsy and Jim; Patty and Constance; and, of course, Nicholas.

This book was written while I was on sabbatical from the University of Washington. Much of my inspiration over the past decade has come from the people of that marvelous institution—the best public university in the world. Several people read parts of this manuscript: Bill Cannon, Betsy Ward, Steve Ward, Bruce Crowley, Don Grayson, Judith Chandler, and Gordon Orians. Sue Bolssen did her usual superb job in production. Brian Tart and Leslie Meredith were the editorial midwives (sorry, Brian) who metamorphosed manuscript to book. Surely errors of fact have crept in; for these, I apologize. I did my best.

Contents

Introduction

I

Regulator in, air's on, hold the mask, clear below, go.

The small boat lurches violently as I roll backward over the side, the large SCUBA tank strapped to my back cushioning my fall into the warm ocean. Finding "up" after a moment's disorientation, I look about while awaiting my companion's entry. The water is crystal clear, typical of the coral reef regions of our world; I feel that I can reach out and touch the seabottom, although it is at least twenty feet below. A noisy splash heralds the arrival of my diving buddy, and as he swims toward me I start downward. My equipment on this dive seems incongruous: Although the tank, mask, and fins fit like old familiar clothes, the rock hammer and chisels tied to my weight belt feel annoyingly out of place. But I will need them today, on a seabottom beneath the Tanon Strait in the southern Philippine Islands.

We reach the bottom and then follow its gently inclined slope toward deeper water. The ocean around us is clear, bright, and warm as we power over a forest of gaily colored sea fans. The gentle swell of the surface can barely be felt here, but it is sufficient to cause the large, reticulated creatures slowly to wave in unison as we pass. For me it should be a familiar journey, for I have dived in myriad coral reefs scattered across the tropical oceans of our world. But this dive, my first in the Philippine archipelago, is a unique experience, and an unsettling one. The sea-fan forest is indeed beautiful, but its presence here is inconsonant with my past experiences in warm tropical seas; this shallow bottom should be the site of a thriving coral reef, where hundreds of coral species lie meshed together in a lithic framework of diversity, a place where the variety of fish, plants, and invertebrate animals coexisting with the corals should be measured in the tens of thousands. But this bottom is devoid of the giant coral cities, and although gaudily festooned with the sea fans and a few other cnidarian relatives of the stony corals, it appears, on closer inspection, to be curiously barren, like a ghost town emblazoned with flapping flags and banners but empty nevertheless.

We find a gully carved into the rocky bottom, and follow it downward. I pump more air into my buoyancy compensator as we descend to sixty feet, coming, finally, to a giant underwater cliff. I feel a moment of vertigo at the cliff's edge, for it drops suddenly downward into a deep indigo blue. Far below I can barely discern a steep, rubbly bottom at the cliff's base, last stop before the inexorable plunge toward the basin of the deep Pacific Ocean. I float out over this blue chasm and begin to sink downward until a far-off movement along the reef wall catches my eye. A white-tipped reef shark slowly emerges from the gloom, attracted perhaps by our strange, noisy bubbles, and surely hoping to find a meal. It circles once and then returns to its solitary patrol of the reef scarp. As the shark disappears into the distance I am once again struck by the strangeness of this place, and half expect to hear Rod Serling's disembodied voice tell me that I am indeed in the Twilight Zone. The sides of this giant underwater wall should be home to numberless fish; elsewhere in the warm regions of the tropical Pacific such habitats are like busy streets on market day, thronged with piscine shoppers and gawkers, children and aged, a crowded diversity all engaged in disparate endeavor. But here, in warm, clear water, I stare outward into a nearly empty sea.

We turn from the steep wall and retrace our path back into the brighter, warmer shallows, until once again I am surrounded by swaying sea fans. I settle onto the bottom and, using my rock hammer, begin to pry away the carbonate rubble. The surface material is composed largely of dead sea-fan stems and skeletal remains of the other creatures living among them. But as I dig deeper I come across different shards, the fragmented remains of corals and shells no longer living at this locality. Many are highly eroded and worn, and this is no surprise, for all carbonate material in the tropics is soon invaded and ultimately disintegrated by myriad species of algae, worms, and sponges, all boring into limy substrata in search of food or shelter, eventually destroying them. Other material I find, however, seems fresh and little invaded by the borers but is highly fragmented nevertheless. I begin to find the broken skeletons of species after species of coral, all typical of the tropical Indo-Pacific region, all from corals that I had expected to find alive here, not dead. With my air running out I finally finish digging a shallow trench several feet long, revealing a three-part stratigraphy. The uppermost, youngest layer is composed of skeletal remains from the few corals and sea fans now living on this bottom. The lowermost layer is completely different: It is packed with large, mainly intact skeletons of a diverse coral reef assemblage, most still in living position. It is the remains of a giant coral reef, and stretches downward into the seabed as far as I can dig; I have no doubt that this coral-rich bed is hundreds or even thousands of feet thick, the geological record of a once-thriving and stable community that had lived for thousands of years here, on this Philippine seabottom so perfectly sited to sustain the coral reefs that characterize the warm shallow oceans of our earth. Finally, I look at the thin layer sandwiched between these two records of such differing communities. It is composed entirely of the shattered skeletons of coral species, the same species found just beneath. But in this middle layer all of the skeletal fragments are small in size, and some are discolored as if by great heat. The thin, middle layer suggests that some repeated bombardment of the coral communities had occurred, shattering the limy skeletons into a fragmented bed and killing off the delicate reef community of organisms in the process. It is a layer created by catastrophe, and brings to mind other thin rock layers I have seen, layers much older, stratal layers giving testimony to great catastrophic extinctions occurring on the earth millions of years ago. One such layer is found in rocks 250 million years old, and it marks the extinction of the earth's first great fauna. The second such layer is 65 million years old; it is the gravestone of the dinosaurs. But the layer I have just unearthed isn't millions, or thousands, or even hundreds of years old; no, quite the contrary: It was produced very recently. Tired, and sick at heart, I begin my ascent toward the light and warmth of my world, for once happy to leave the crystalline waters of a tropical sea, for I have just seen evidence of an extinction. Thirty years ago the bottom beneath me was home to a teeming coral reef community, and the seas around this storehouse of marine diversity were alive with fish. But then a new type of fishing was introduced and condoned by the government of Ferdinand Marcos: dynamite fishing. Sticks of explosive thrown onto the reefs killed hundreds of fish at a blast, creating a harvest of bodies floating to the surface that had only to be scooped from the sea. A golden age had seemingly begun for the Filipino fishermen. But these men did not recognize until too late that the reefs they bombed were breeding as well as feeding grounds for not only the reef dwellers, but for most fish in the region. Dynamite quickly reduced the reefs to rubble, and most food fishes began to disappear. Within two decades the fishing yields in the Philippines dropped by well over half, while the population of humans, most of whom depended on fish as the main source of protein, almost doubled. Belatedly the government tried to halt the practice of dynamite fishing, hoping for a quick fix. But corals are slow-growing creatures; even if left alone, it will take many centuries to restore the Philippine reefs to their previous lushness. Yet the reefs will not be left alone. The crater-strewn, underwater battlefields of the Philippines still harbor occasional fish large enough to eat, and the hungry fishermen have more mouths to feed every day: The current human birthrate in the Philippines doubles the population every twenty-six years.

I once saw the barren remains of a field after an invasion of locusts. With this unbidden memory in mind I climbed from the Philippine sea.

2

Two weeks have passed since my first dive into the clear, dead waters of the Philippine seas. The stars lie over me in shimmering splendor; they dance, change color, mocking my attempts at sleep. A cacophony of sound also conspires against me: The gentle lapping of waves on the nearby shore, the busy clicking, rasping, thrumming, and throbbing of insect armies on the small island around me, even the gentle snoring of my nearby companions drives away any possibility of sleep. Perspiration rises from my forehead; the heat of this torpid night is the most unwelcome of guests, visible in its stranglehold on the once-unblinking stars. I roll over as the persistent high-pitched whining of another hungry mosquito draws my attention; once again I rue my sleeping arrangements. My companions and I are lying on a beach, without mats or blankets. Not that we need warmth in these stifling tropics. But I find that a sheet is a very comforting thing to have, even on a very hot night. Despite knowing that all of mankind's ancient predators are either extinct or in zoos, it still seems some slight protection against the myriad demons that have prowled just outside human campfires for most of the last million years, and patrol there still just beneath the logical layers of my mind.

Turning restlessly, I can see the nearby village, marked by several low fires; all else, however, is stygian blackness, the impenetrable darkness of the equator. I wonder at the lack of movement or noise coming from the ramshackle clutter of huts, for several hundred people are crammed into the village, yet they give little sign of life. People on low-protein diets apparently don't stay awake long after the sun sets; an active nightlife is a relatively new thing for humanity, and applies only to that portion of us carrying a full belly of food when darkness comes.

My presence on this small island, no more than a quarter mile across, was unplanned, a last-gasp effort to conduct a scientific experiment. I had journeyed to the Philippines to trap specimens of the chambered nautilus, last living remnant of a once-great Mesozoic fauna, but every effort up till now had failed. My host in this endeavor, Dr. Angel Alcala, brought me to this island only after all of our efforts in more traditional nautilus fishing grounds had proven futile. Nautiluses have been trapped by the Filipinos of the Tanon Strait area for generations, their flesh eaten, their shells turned into buttons. But the once-prolific waters of the Tanon Strait between Negros and Cebu in the southern Philippines is now a dead sea; gone are the huge schools of fish, overexploited, their hatching grounds destroyed by two decades of dynamite fishing; gone are the abundant stocks of nautilus, their once-rich feeding grounds poisoned by heavy-metal dumping and siltation; gone too are the seabirds, their eggs and young destroyed by DDT concentrations running off the rice paddies. Alcala had managed to save this tiny island from the ravages of the dynamite fishing that had destroyed virtually every reef once lining the Tanon Strait and southern Negros. He had brought me here to see this last bit of undisturbed southern Philippine reef, but had been dismayed to find a new village now sitting on the island, a community without electricity or running water. The enclave of fishermen had been driven here by hunger, for only the reefs held harvestable fish, and this was the only reef for miles around. During the daylight we watched as fishermen rowed in their dugouts past the signs proclaiming the limit of the reserve, dropping their lines and nets into the aquamarine reef waters. Alcala said little; when your children are hungry, you do what you must to feed them.

During my two weeks on Negros I had been amazed at the sea of humanity lining the shores. Small huts paved the lowlands, stopping only at the high-tide line of the sea. The roads and beaches were awash in children; crying, laughing, playing as all children do, but children tiny and bony, children hungry in a land that once boasted an abundance of food. That wealth of food is now largely gone; the rich highland rain forests once lining the island sides have been logged and exported to far shores. The once-lush lowland forests have been burned, replaced by rice fields on now-exhausted soil. The loss of the forests brought about the disappearance of native animals and plants that had served as food for the Filipinos for thousands of years. The ancient, twin horns of cornucopia, the Philippine forest and sea, are now empty places of little diversity, places filled only with weeds. I thought of the blasted remains of the coral reef I had seen on my first dive in the Philippines and remembered a lesson I had learned in graduate school: Mass extinctions, times of greatly elevated rates of species death, have always been most prolonged, pronounced, and devastating in the tropical regions of the earth. But my professors had been speaking of past mass extinctions, never dreaming even two decades ago what 5 billion hungry humans are capable of doing to the earth's biota.

3

Mass extinctions are relatively short intervals of intense species death. During the last 570 million years of earth history, the time since the advent of skeletonized creatures on the earth, there have been about fifteen mass extinctions. Five of these may have involved as many as 50 percent of the earth's species, and two can be classified as "major," in the sense that they completely reorganized the ecosystems in the sea and, more relevant to humanity, on land.

The first of these two major mass extinctions occurred 245 million years ago. Being the oldest, this first event is still the most poorly known, and its causes are largely unresolved. Many earth scientists believe that it was brought about by a slow yet inexorable change in climate and sea level occurring when forces of continental drift caused the earth's great continents to merge together slowly into a single, gigantic supercontinent. This was a new world of endless glaciers and waterless deserts, of unearthly temperature extremes between summer and winter: a land of extinction. By the time the continents had finally separated from their lethal tectonic embrace, more than 90 percent of the earth's species had died. This great extinction swept away most of the marine and land-living animal life, ending a 200-million-year-long evolutionary history that geologists have named the Paleozoic Era.

The second great mass extinction took place 65 million years ago. Like its predecessor, the second event was caused by several factors, including climate changes and a sudden change in sea level. But the culmination of this extinction, and by far its most dramatic element, took place when one or more large asteroids or comets crashed into the surface of the earth, collisions whose violence ended the 150-million-year-long Mesozoic Era, the Age of Dinosaurs. These titanic impacts produced a fiery hell of burning forests over much of the earth's surface, accompanied by giant tidal waves and great volumes of poisonous gas: These were the short-term effects of the first few hours and days. But even more lethal may have been the months of darkness that enveloped the planet after the impacts, as unnumbered tons of earth and extraterrestrial debris initially blasted upward by the collisions slowly sifted out of orbit, all the while obscuring the sun in an endless night, a plant-killing night. On land, and even more so in the oceans, the plants died—and with the death of the plants came the starvation of so many creatures that fed upon them. Well over 50 percent of all species on the earth perished. Those that survived—crawling from holes and burrows, or emerging from the depths of the sea—began life anew in a much-emptied world.

Following the death of the Mesozoic world, a new fauna inherited the earth. In the 65 million years since the last of the dinosaurs died in fire and brimstone, the survivors and their descendants have multiplied to levels of diversity unseen during the previous two eras of life. Sadly, there is now mounting evidence that a third great extinction episode has commenced, a rising tide of death that will end the third great era in the history of life on the earth, the Cenozoic Era, the Age of Mammals. Like the previous two events, the current mass extinction has a complex history and no single cause; like the previous two, it is not a sudden event, an unheralded Armageddon of little duration. The current mass extinction has been unfolding for millennia, and unlike the greenhouse effect, global warming, or the hole in the ozone, it is visible without sophisticated imagery or complex computer modeling. It is real, and it is happening to a greater or lesser degree all over the globe; it is most apparent, however, in the tropics. It will not eliminate life from the earth: No mass extinction does that. But enough species will die that the nature of life on the earth will be forever changed. It can be called the Third Event.

Many scientists dispute whether an extinction is currently taking place at all, or suggest that we are facing the prospect but have not yet begun the experience. Others agree that we are indeed in a period of increased extinction, but that the net result will little change the earth's flora and fauna. I do not share such a sanguine view. I believe that the Third Event is well under way, having started with the dawn of the Ice Age, about 2.5 million years ago, and since then accelerating in its rate of species destruction. In some ways it is very much like the dinosaur-killing Second Event of 65 million years ago, when a biosphere already stressed by rapid changes in climate and sea level was knocked into mass extinction by the impact of asteroids, striking, according to new evidence, simultaneously in North and Central America. A very similar scenario is currently unfolding. Over 2 million years ago, giant glaciers began to cover large portions of the earth, changing climate and sea level on a global scale in the process. And then, 100,000 years ago, another great asteroid hit the earth, this time in Africa. That asteroid is named *Homo sapiens.*

All species evolve until they die. Extinction is the end of evolution.

Part One

The First Event

Chapter One

The Cape

I

A stiff wind is blowing in from the sea: cool, salty, carrying with it the call of birds. I am standing at the edge of a sheer cliff, a vertical precipice falling more than two hundred feet to crashing surf below. The sides of this monumental scarp are the homes of nesting seabirds. They are oblivious and impervious to the vertical nature of this three-part contact among land, air, and ocean. I look south, out to sea, where endless whitecapped waves stretch to the horizon; the next land in that direction is Antarctica. Far, far to the west lies South America; to the east, Australia. Turning, I look north, and see the western and eastern coasts of Africa, one washed by giant breakers rolling inward from the South Atlantic, the other by the pellucid stillness of the Indian Ocean. This gigantic promontory on which I stand, the Cape of Good Hope, is Land's End for the continent of Africa, final stop for the millions of miles of roads, tracks, jungle trails, and windswept desert paths filling this giant continent, all eventually interconnected, all ending here. Rising hugely out of the water, this monstrous, high crag of rock seems a fitting terminus. For me, however, it is the start of a journey, not the end.

I step back away from the edge of the yawning cliff as a further blast

PHOTO BY THE AUTHOR.

The Cape of Good Hope.

of wind tugs at my clothing. Turning to change lenses in my camera, I find myself looking into the faces of some of the Cape's permanent residents: A troop of baboons has silently gathered behind me. While members of this busy assemblage wander about the rocks, pursuing their various chores, a large female baboon carrying a small infant sits down on a rock and stares at me with level brown eyes, perhaps appraising my ability to fork over some food. The adult baboon is an alien creature to me, not of my species; I find myself feeling little kinship for this fellow primate, but the baby is another matter. It disengages from its mother and makes a lunge at a large black lizard sunning on a nearby rock; missing the lizard by a mile, the young baboon seems human in its disappointment, with body language and facial expression all too familiar to the parent of a human child. And with this contact, my first with a member of the African fauna, it finally hits home that on many levels I am a very long way from my world; the twelve-hour, night-long trip on a packed 747 really did fly me to Cape Town, to Africa, to the birthplace of mammals, mankind, and perhaps dinosaurs as well.

Africa holds many secrets. Locked within the gritty sediments of ancient Africa are hoary mysteries of pharaohs and Numidian kings as well as the bones of past inhabitants much older. Here lie the birthplaces and graves of the first humans, and of creatures far more ancient than humans: Interred within ancient African sandstones are quarter-billion-year-old skeletons from ancestors of the Class Mammalia, of which humanity is but one member. They are buried two hundred miles from where I stand, in a great desert known as the Karroo. It is a rocky badland, like the desert regions of the American West, but much older. The Karroo's ancient, scoured, and scored rocks hold the story of an animal empire that painfully rose out of the swamps, first taking tens and then hundreds of millions of years to evolve from wriggling spineless creatures to fish, and then onto the land; finally to shed all need of water through the evolution of lungs and water-resistant amniote eggs; to cast free the shackles of cold-bloodedness and stand on the threshold of the mammalian evolutionary grade: This great dynasty sat poised to exploit the earth. But at the moment of its greatest triumph the empire of the protomammals was struck by a catastrophe that almost completely destroyed it: The First Event delivered almost all the Karroo fauna into the oblivion of extinction.

The few protomammals surviving this great mass extinction inherited a nearly empty world. But these survivors, only one evolutionary step removed from attaining the status of being the first true mammals, found that they now had to share a world they once dominated with a formidable, new competitor: a small crocodilelike reptile, itself but one evolutionary step short of its destiny, becoming the first true dinosaur. Armed with sharp teeth and the ability to produce a great number of new species in very short order, this creature very quickly sidled over to the surviving protomammals and began to make meals of them. Following the First Event, our ancestors competitively faced the ancestors of the dinosaurs in a winner-take-all fight for land-animal dominance of the Mesozoic world. We lost. Our new overlords, the dinosaurs, then controlled the earth for over 100 million years, and probably would be here still, but for the advent of the Second Event.

Another cold blast of wind brings me back to our world, where one human and a troop of baboons shiver together on a rock high above the sea. Unnumbered seabirds scream overhead: Descendants of the long-dead dinosaurs, they seem to be either laughing or crying at the big and small primates beneath them on this rock; laughing, perhaps, at the mammals' fate following the First Event, or crying for the equally dismal fate of their dinosaurian ancestors, during the Second Event. But perhaps their mournful cries in this antipodean wind are for the fate of their own kind, suffering so cruelly in the current extinction, the Third Event.

2

The shoreline north of the Cape of Good Hope is a ruggedly beautiful alternation of jagged rocky headlands and pristine white beaches stretching to and then past the city of Cape Town, South Africa. At the edge of the city lies a rocky coastline lined by forests of brown kelp and the bobbing heads of otters, surfers, and seals. Along this shoreline, a young naturalist named Charles Darwin stopped in 1834 to view a fascinating geological site. So well known as one of the great biologists of all time, Darwin initially was trained as a geologist. This education served him in good stead during his epic five-year voyage on the *Beagle,* as well as later, when, in the quiet of his study, he had to wrestle with vexing objections that the geological and fossil record seemed to pose to his nascent Theory of Evolution.

HMS Beagle, carrying Darwin, naturally stopped in Cape Town during its long trek around the globe, for at that time this scenic city was known as the Tavern of the Seas; it was a necessary watering station for the merchant fleets (and their thirsty sailors) plying the waters between the rich spice islands of Asia and the home ports of Europe. During his stay in Cape Town, Darwin made many geological excursions, among them a visit to a narrow beach where a succession of dark sedimentary rocks known as the Malmsbury shales lie in contorted stripes; a century later the city fathers of Cape Town erected a stone marker commemorating the visit. One hundred fifty years after Darwin I descended onto this same beach, to follow a lowering tide. (There is no news about a marker for me, but one can hope.) Newly exposed sandstones and shales covered by a rich fauna of intertidal life gleamed wetly under bright sun on this day. But teeming with life as the surface of these rocks are, internally they seemed dead, for they contained no fossils whatsoever. They are barren strata, and not because they were formed in environments where life could not live; the strata on this Cape Town beach were deposited on a seabottom almost a billion years ago, during a time when life was represented mainly by scums of algae and drifting one-celled infusoria. Life, so rich on our planet today, is almost as old as the earth itself. But for nearly three-fourths of that four-and-a-half-billion-year history, life on earth was simple and dull, and left little or no fossil record; at best, we have the traces of one-celled creatures locked in cherts, or fossilized mounds of blue-green algae from shallow oceanic shorelines or freshwater lakes. It was not until about 700 million years ago that life began to try rich new experiments in design, freeing itself forever from the shackles of single-celled body plans. Thus began a great surge of diversification, when the simple plants of that long-ago time began to be joined—and eaten—by waves of newly evolved creatures, a tide of new life sweeping in to fill the seas with multicellular animals and plants. The metazoans, or multicellular life of which our species is but one example, were a late arrival in the earth's history. But once unleashed, the armies of metazoan creatures quickly multiplied and diversified. Around 600 million years ago these creatures developed another innovation: They evolved skeletons and hard parts. Partly serving as protection, partly as a means to attach musculature and thus allow new ways of feeding and locomotion, the advent of skeletons made these ancient creatures candidates for preservation as fossils, and changed the nature of the fossil record in the process. This transition marks the boundary between two great eras: the Precambrian Era and the succeeding Paleozoic Era.

The boundary between these units was long thought to be sharp, for apparently unfossiliferous strata in many parts of the globe are overlain by strata rich in shelly fossils. This supposedly sudden transition from a lifeless world to one packed with skeletonized creatures baffled both Darwin and later geologists alike, for the apparently instantaneous appearance of skeletonized life in the fossil record seemed to contradict Darwin's Theory of Evolution. Darwin maintained that new species arose as a gradual process, not instantaneously. If his theory was correct, the skeletonized fossils marking the base of the Paleozoic Era must have been the products of a long evolutionary period prior to their appearance. But this did not seem to be, for Darwin and his contemporaries could find no fossils other than simple plant forms in strata underlying the shelly faunas; indeed, the apparently sudden appearance of life, at a time we now know to have occurred some 570 million years ago, became one of the most potent bits of ammunition in the arsenal of Darwin's many critics. Darwin, in his lifetime, was never able to refute this stringent criticism of his theory. Only much later have paleontologists proven Darwin correct, for strata beneath those containing the supposed first appearance of fossil skeletons do indeed contain the ancestors he theorized and sought after. They were long overlooked or missed, however, because of their very small size; most fossils from this period of life, known as the Tommotian Period, had skeletons very hard to detect unless special processing techniques are used to extract them from their entombing matrix. Such methods had not yet been dreamed of by Darwin and his contemporaries. These Tommotian microfossils seem to have appeared about 10 million years prior to larger skeletonized life, and they themselves must have an even longer family tree of ancestors without any skeletons capable of fossilization. The supposedly "sudden" appearance of skeletonized life 570 million years ago is simply the first appearance of creatures with large skeletons, producing fossils easily noticed. We now have a worldwide record of a diverse fauna with fossilized skeletons no more than a millimeter or two long,

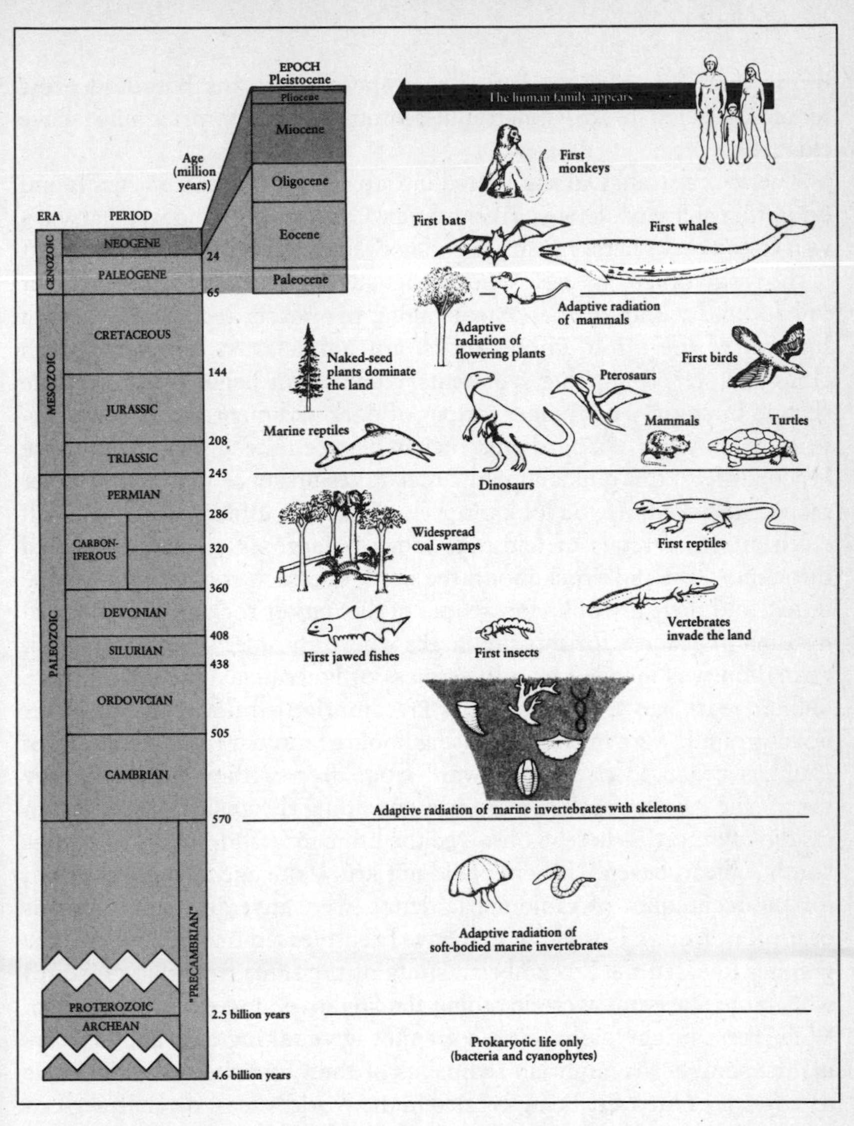

Geological Time Scale, discussing the time units used in this book.

bearing out Darwin's prediction that ancestors to the basal Paleozoic fossils collected by his nineteenth-century contemporaries must have existed.

The rocky strata Darwin visited in Cape Town contain no fossils and do not have a well-defined Precambrian-Cambrian boundary; Darwin's visit to the rocks here, I suspect, had nothing to do with his later interest in the formation of species. What intrigued him was a geological contact to be found within the dark strata along the beach, recording a case of survival of the fittest among rocks, not animals. As you walk south along the steeply dipping sedimentary beds, you begin to see a subtle change in color. The thinner stripes of dark sediment first show a few scattered patches of a hard pink rock in between them; farther down the beach more of this pink and white rock takes up space within the darker sedimentary beds. If you look more closely at this lithic invader, you will see beautiful crystals of feldspar, some as large as an inch, scattered throughout the pink rock. Soon the hard, speckled rock dominates the beach, and there are but a few stripes of the darker rock left, surrounded by a sea of granite, for this site marks where an older sedimentary rock formation was invaded by a huge mass of igneous magma. About 570 million years ago these dark, cold Precambrian strata, then buried far underground, were invaded by rising pools of molten rock. As these hot magmas began to rise surfaceward from deep within the earth, they melted the country rock in great heat, invading the ancient rocks like an insidious cancer. When he observed the lithic invasion preserved on this South African beach, Darwin could not know the age of these granites, for the techniques of radiometric dating were invented long after his death. But I think Darwin would have been pleased had he known: The granites invaded the Precambrian strata at the same time that creatures with large skeletons were invading the shallows and deeps of the sea. While these bright, gaily colored granites were taking over and destroying the somber Precambrian sediments of the Cape peninsula, fantastic new forms of life were being created in the world's seas, diversifying in a mad rush never seen before that time, and leaving a fossil record that marks one of the most obvious stratigraphic boundaries on earth: the start of the Paleozoic Era, the time of skeletonized life. Within a period of several million years, the first trilobites crawled and swam in the warm sunlit waters; clamlike brachiopods festooned the seabottoms, while flotillas and fleets of worms, jellyfish, sponges, and other newly evolved species began to crowd the seas. The invasion of the bright Cape granites into the lifeless Precambrian Malmsbury shales seems a pleasing metaphor for this transition; like the great diversification of life that marked the start of the Paleozoic Era, the granites infiltrated and ultimately overwhelmed the ancient setting of this shoreline, leaving testimony of a great revolution for all to see.

3

Charles Darwin's visit to Cape Town took place near the end of his journey around the globe; after leaving South Africa, the *Beagle* headed up the coast, then west to South America, finally landing in England some months later. Darwin's notebooks were filled with records and observations of his trip, and he carried within them, and his head, the seeds of a scientific revolution still unfolding and still controversial. The theory of species formation through a process of natural selection is one of the great scientific breakthroughs in the history of science, a theory so revolutionary that it changed not only the way in which scientists looked at nature but the way mankind looked at itself.

Following his voyage, Darwin did not hurry home to his study and write up his theory of evolution; quite the contrary. He reflected for years as the idea took root and grew; he worked, writing meticulous monographs about the various creatures he had seen, but all the while contemplating concepts larger than the taxonomy of barnacles or the nature of earthworms observed on his long voyage. He was seemingly reluctant to publish his great theory, however, and only when another biologist independently arrived at conclusions eerily similar to his own did Darwin finally put pencil to paper. His monumental work, *On the Origin of Species,* was published in 1859, more than twenty years after his return from circling the globe in the tiny *Beagle.*

The same twenty-year period when Darwin contemplated how new species form was a momentous one in the history of science. During this time Darwin's zoological colleagues were busily cataloging and naming the animals and plants of the world, using Carolus Linnaeus's method of binomial nomenclature, where each separate creature received a unique genus and species name. Darwin's geological colleagues used the same methodology in describing the numerous fossils they were finding across Europe. In the process, they erected a system of geological time still in use today. Perhaps more than anything else, the pioneering geologists of the early and mid-nineteenth century began to realize the immense antiquity of the earth. This realization, which certainly was communicated to the scientifically literate Darwin, gave credence to a theory itself requiring enormous periods of time to be operational.

4

In the early nineteenth century the industrial revolution sweeping Europe gathered an inexorable momentum. Roads were built, canals cut, factories opened; railroads began to cross the landscape as trade flowed freely across the continent.

All of this took raw materials and energy. Giant quarries were opened to yield sufficient building stone and lime for cement; forests were cut to provide lumber for dwellings and businesses. But perhaps most pressing of all was the search for the two main materials necessary to run the great engines of this upheaval: metal and fuel.

From these needs the science of geology was born. Although academic interest in the history of the earth certainly existed prior to the nineteenth century, the search for raw materials and fuel created a need for geologists, for only the earth itself could yield these two requirements.

The pioneering geologists soon discovered how complicated the surface of the earth is. The untold eons of mountain building and destruction, the rise and fall of the seas, the compaction, pressure, heating, cooling, sedimentation, erosion, and myriad other processes shaping the surface of the earth have created a crust almost incomprehensible in its complexity. From within this labyrinth, the sought-after treasures of metal, stone, coal, and oil must be extracted, and to find them geologists must first decipher and interpret the complex geological history of the earth's surface. History—geological history—provides the key to these mineralogical fortunes; history, which itself is but another manifestation of time. The key to deciphering the earth's complex history was a reliable method of dating rocks.

Telling the age of a rock turned out to be very difficult indeed. The earliest attempts centered on the relative hardness and nature of rock.

Igneous rocks, such as the speckled granites making up the cores of mountains, seemed most ancient of all, while the layered, or sedimentary, rocks seemed softer and, perhaps, younger. But how to tell the relative ages of the sedimentary rocks themselves? Since coal, oil, and even many mineral deposits are contained within sedimentary rock successions, or assemblages, it became imperative for the pioneering nineteenth-century geologists to devise some method of telling the relative age of these rocks scattered across the globe. Within a given succession of sedimentary rocks, such determination is relatively straightforward; since the time of the great seventeenth-century scientist and philosopher Nicholas Steno, it was known that sedimentary rocks were deposited in a superpositional order, with older layers found at the bottom of the pile. Thus, if the superpositional order of sedimentary rocks in different areas could be determined, the relative ages could be ascertained. But therein lies the trick.

In some places on the earth, great successions of strata lie relatively undisturbed. The Grand Canyon is one such place, where huge piles of strata, relatively unchanged for more than 500 million years, lie in their original, horizontal orientation. By climbing up through this succession one is literally going up through time, bed by bed. Unfortunately, such undisturbed piles of strata are relatively rare; all too often the long and tortured history of the earth's surface has thrown the originally horizontal strata into broken and contorted mazes of tilted rock, making age determinations based on superpositional relationships difficult or impossible.

By the early nineteenth century it was clear that a successful search for minerals and coal would require an accurate way of dating sedimentary rocks. But how? The composition of various rock bodies found on the surface of the earth was, in itself, an unreliable clue, for even the earliest geologists soon realized that rock type per se was independent of time; a sandstone forming today can look identical to sandstones formed 500 million years ago. Clearly, a timekeeper independent of lithology was necessary.

The answer to this dilemma was found by an English surveyor named William Smith. Smith was involved in canal building in various parts of Britain. A keen observer, he noticed that although the succession of rock types was often quite different from district to district, the succession of fossils found within given units of strata was often the same. Smith had stumbled on a great discovery: Fossils, rather than rock type, could provide a way to determine the age of sedimentary rocks.

Smith's late-eighteenth-century discovery revolutionized geology. (It was first announced in a pub, appropriately enough—most geologists have been, currently are, and probably always will be enthusiastic patrons of beer.) Because sedimentary rocks of the same age, even if from widely separated locales, often contain the same groups of fossils, geologists could demonstrate contemporaneity of formation—the first and most crucial step in unraveling the geological history of a given region. Within two decades of Smith's discovery, which came to be known as the principle of faunal succession, this method was being applied to strata in many parts of the world, while in universities an entirely new discipline was born: paleontology. The study of fossils became a necessity as the need for coal and then oil grew ever more important to an energy-hungry economy, for it was the fossils that gave the ages of the rocks.

With the aid of fossils, geologists soon began to subdivide the earth's sedimentary strata into large-scale units. These subdivisions, although originally based on actual rock bodies, became de facto units of time. For instance, an English geologist named Adam Sedgwick spent several summers in the early part of the nineteenth century studying strata found in Wales. These rocks showed the transition between unfossiliferous strata below and strata filled with fossils above, the transition we now know to mark the start of the Paleozoic Era. Sedgwick named the fossiliferous rocks the Cambrian System and used, in his definition of this group of rock, the characteristic fossils enclosed and found within these strata. The Cambrian Period was defined as the block of time during which these strata were deposited and the fossils found within them were actually alive. We now know that this unit of time started about 570 million years ago and ended about 500 million years ago. Although Sedgwick's strata are found only in a restricted part of Wales, we refer to all rocks on the earth as belonging to the Cambrian System if it can be demonstrated, through fossil content or some other means, such as radiometric age determination, that they were formed between 570 and 500 million years ago. (The ages in years now assigned to various geological time units is a twentieth-century advance. Such determinations are made by measuring minute quantities of naturally occurring isotopes within rocks, using an instrument known as a mass

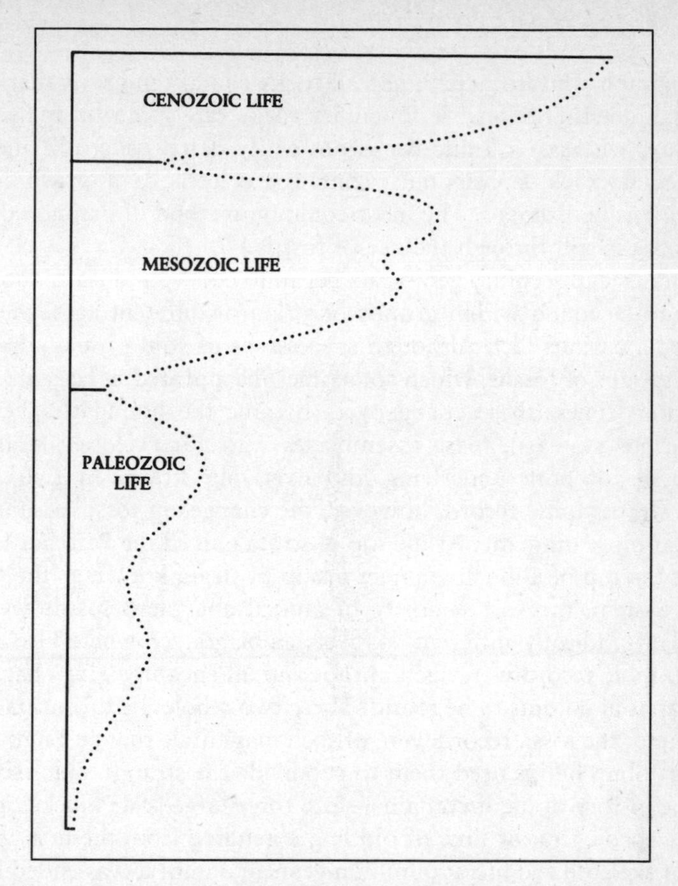

John Phillips's diagram of diversity of life through time, printed in 1860. This figure shows not only the timing and relative severity of the major mass extinctions, but was an accurate model for the progressively increasing diversity of life through time. It was an incredibly prescient model for the time.

spectrograph. Unfortunately, not all rocks contain minerals that allow such age determinations. Sedimentary rocks rarely can be analyzed in this way, whereas volcanic rocks are easily dated using this method. Sedimentary rock deposits must contain a volcanic lava or ash flow to be dated using isotopes. The most common method of dating sedimentary rocks is still through the use of fossils.)

As nineteenth-century geologists began to collect and learn about the fossils to be found within groups of strata of different ages, they discovered a curious fact. Although successions of strata usually had different groups of fossils, which sometimes disappeared in large numbers at various times (these changeovers became the boundaries between the various systems), these assemblages were often closely related to the fossils of both underlying and overlying strata. At two places in the stratigraphic record, however, the changes in fossil assemblages were far more dramatic. At the top of strata named the Permian System and at the top of a much younger group of strata known as the Cretaceous System, the vast majority of animal and plant fossils were replaced by radically different fossil assemblages. Nowhere else in the stratigraphic record were such abrupt and all-encompassing changes in the faunas and floras to be found. These two wholesale turnovers in the makeup of the fossil record were of such magnitude that an Englishman named John Phillips used them to subdivide the stratigraphic record—and the history of life it contains—into three large-scale blocks of time: The Paleozoic Era, or time of old life, extending from the first appearance of skeletonized life 570 million years ago until it was ended by the gigantic extinction of 250 million years ago known as the First Event; the Mesozoic Era, or time of middle life, beginning immediately after the great Paleozoic extinction and ending 65 million years ago with the Second Event; and the Cenozoic Era, or time of new life, extending from the last great mass extinction to the present day. Phillips suffered no misapprehensions about the cause of these three divisions: He realized that at two times in the earth's past, life had almost been extinguished, with only a tiny fraction of species surviving the catastrophes, whatever their cause. Phillips concluded that the history of life on the earth had, in the past, been interrupted by mass death—times of wholesale destruction of species and individuals that we now call mass extinctions.

At the time of Phillips's work, the concept that a species could go extinct was still quite new. Although many of the animal and plant fossils found from the fossil record of that time looked quite unlike creatures living then, how could scientists be sure that the bizarre creatures were not still living in some as yet unexplored corner of the world? This notion, whose last gasp was manifested in Arthur Conan Doyle's wonderful evocation of Mesozoic life, *The Lost World,* seemed a distinct possibility to late-eighteenth-century naturalists. A strong religious bias was also at work: If God, in His wisdom, went to the trouble of creating a given animal species, why would He remove such a perfection later? Surely all species ever created had to be still alive somewhere. The killjoy who finally burst this bubble was the great French anatomist Baron Georges Cuvier, who demonstrated to almost everyone's satisfaction (then, as now, many religious zealots refused to accept scientific proof) that the fossil bones of large mammals such as mammoths and mastodons are from creatures no longer living on the earth. Cuvier also should be credited with being the first to recognize the two great mass extinctions that John Phillips later used to subdivide the geological record.

Any mass extinction has two parts: mass death followed by the earth's repopulation with a largely new suite of creatures. Cuvier's proof of the reality of extinction solved the first part of the puzzle. But what of the repopulation? The origin of species remained an abominable mystery to most scientists of the early nineteenth century; many, such as the pioneering French geologist Alcide d'Orbigny, were content to leave both the extinction and repopulation to God. Others were not so sure. One such skeptic was Charles Darwin, whose great work gave Phillips the key to the second part of the puzzle: The repopulation of the earth occurred through speciation. The survivors of extinction, finding themselves in a world depleted of other species and with many ecological niches left empty, usually produced new suites of species rapidly.

John Phillips's 1860 summation of the fossil record, and his recognition of the two great punctuation marks inserted into the history of life by the mass extinctions occurring at the end of his newly defined Paleozoic and Mesozoic eras, was of importance not only because of his recognition of the severity of these two great extinctions. His paper also marked the first serious attempt at estimating the diversity, or number, of species present on the earth during the past. Although Phillips made no attempt to arrive at actual numbers, he did produce a drawing showing his estimation of the relative number of species through time and of the relative severity not only of the two great mass extinctions but of other, lesser reductions in species numbers also observable in the stratigraphic record. Phillips thus portrayed graphic evidence of mass extinctions as well as the first estimate of diversity through time—from the Cambrian to the present—in prescient detail.

It is a testament to the severity of the two great mass extinctions that they were recognized so soon after geologists had begun studying fossils. But if the identification of two major crises in the history of life came as a surprise, the aftermath of each of these events may have been even more so. The fossil record seemed to show that the times after mass extinction were periods of little life on the earth (at least in terms of the number of species). Over time, however, the diversity of fossils began to increase, until the diversity levels prior to the extinctions were reached. But Phillips noted that diversity levels did not stop there: Instead, they continued to climb, until levels far in excess of the previous times had been reached. The implications were clear: The mass extinctions cleared out old communities, allowing a rediversification of creatures vastly different from those of before. However, the mass extinctions did not only change the biotic makeup of the earth; in some way they seemed to make room for larger numbers of species than were present prior to the extinction. Far more creatures were present in the Mesozoic than the Paleozoic, and then far more again in the Cenozoic. But were these figures accurate, or were they simply an artifact of the fossil record? Phillips's diversity diagram showed species richness in the past not as actual numbers of species but simply as a number relative to present-day diversity; how could such a graph be taken seriously when no living scientist had any idea how many species actually existed on the earth at that time?

5

How many different kinds of species are there on the earth today? The question, one of the fundamental issues of modern biology, has attracted the attention of scientists for several centuries. It is also one of the most pressing questions facing our global civilization. If, as I suspect, we are in the middle of a mass extinction as profound as any that can be recognized in the geological record, we must have some reliable estimate of species diversity, if only to know the rate at which species are disappearing by extinction. But arriving at such an estimate is fraught with difficulty, and in spite of ever-increasing study on just this issue, the answers—estimates, really—are paradoxically becoming more and more unclear.

Biologists use several terms to discuss the number of species. Diversity is one, and in a sense diversity is synonymous with the number of species present in the environment being studied, be it a pond, a lake, an ocean, or the entire world. But diversity is also a highly technical term, with a rigid mathematical definition; it encompasses not only the number of species, but also the relative abundance of the various coexisting species. For example, two lakes may each contain one hundred species; in one lake, however, the number of individuals belonging to each species is equal, whereas in the second lake, 99 percent of all individuals belong to only one species, with the remaining ninety-nine species represented by just a few individuals each. Even though the number of species in each lake is the same, the diversity in these two systems is very different. Ecologists studying these two lakes would say that the first lake, with an evenly partitioned number of individuals for each species, has a higher diversity than the second lake. In just such a way we could envisage a world where humans put all endangered species in refuges or zoos. Even though no extinctions take place, the diversity of the planet has lessened simply by the relative reduction in numbers of the endangered species relative to the more abundant, weedy species—such as mankind, the consummate weed.

But such a scenario is ludicrous—species *are* becoming extinct. The black irony is that species may be going extinct faster than systematic biologists can describe the new ones they find. At this time, about 1.5 million species have been formally named. Until several years ago, biology texts would tell you that the number of species living today may be as high as 3 to 5 million but is probably less. Thanks to recent work in the tropical rain forests, however, some biologists such as Paul Ehrlich and Edward O. Wilson now believe that we have just barely begun to catalog the earth's present inventory of species. These two respected members of the National Academy of Science have estimated that there may be as many as 50 million species on the earth today, with the vast majority packed into several habitats: tropical rain forests, coral reefs, and perhaps the deep sea. Such complex and, in many cases, far-flung environments are very difficult to census.

How many species have there been in the past? Is present-day diversity higher than at any time in the past, as John Phillips thought, or has diversity been essentially constant back through time? Difficult as they are, the problems in measuring diversity in present-day ecosystems pale compared to arriving at similar measures for past environments. To study the diversity of the past, we must base our sampling on fossils. But fossils are almost never the complete, mummified remains of a creature; usually a fossil is an incomplete representation, normally some skeletal hard part. And most vexing of all, in all probability most creatures that lived on the earth in the past left no fossil record of any kind, since the majority now here have no hard parts capable of fossilizing, and we have no reason to believe that the situation was any different in the past. Worms, jellyfish, slugs, most insects, spiders, mushrooms, moss—these and countless other creatures have few or no body parts that enter the fossil record. They have been preserved as fossils, if at all, only under the most extraordinary of circumstances, such as being caught up in amber or becoming mummified in an oxygen-free environment. Recent studies on modern marine communities have shown that fewer than 30 percent of species have any chance of entering the rock record as fossils. Perhaps in past times a greater percentage of creatures had skeletons, but surely not a significantly higher number; perhaps even fewer than in today's oceans, where hundreds of millions of years of evolutionary struggle have honed and modernized the skeletons and soft parts so necessary to compete in today's world and to fend off today's efficient predators.

With these difficulties in mind, why would anyone attempt to estimate past and present-day diversity? The answer is that it is one of the most fundamental and fascinating questions posed by evolutionary biology. Following the great diversification of creatures in the earliest Cambrian Period, did the number of species increase through time and then level off after a constant number was reached, to remain similar over time? Or has there been an ever-increasing number of species? This question is of utmost importance, not only in understanding the history of past life but in understanding the present-day world as well. Can our earth hold only so many species and individuals, or can their numbers constantly increase? And if there is some inherent limit to the number of species that the earth can hold, has that limit changed through time?

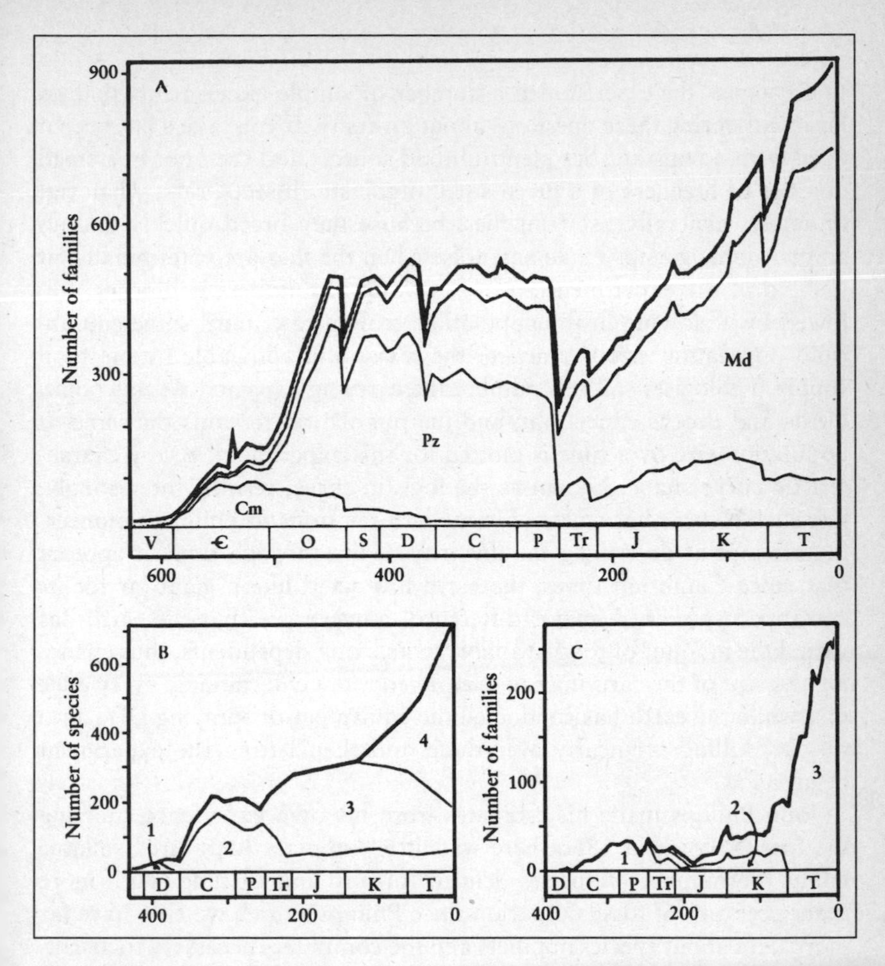

The diversity of life through time (diversity curves) as understood at present. A is based on families of marine organisms, B represents terrestrial plant species through time, and C represents terrestrial, land-living vertebrates. In each, it can be seen that an initial diversification during the Paleozoic Era was followed by a relatively constant diversity. This diversity, in turn, was followed by a great increase in the variety of life during the Mesozoic and Cenozoic eras. Modified from Sepkoski, 1984.

Ecologists have performed a number of simple experiments that go far in answering these questions about diversity. If you take a jar, keep it filled with a constant but plentiful food source, and then put in a small number of breeders of a given small organism (insects, rats, whatever; ecologists generally use fruit flies, because they breed quickly and no animal rights groups make any noise when the flies are exterminated at the end of the experiment), a huge run-up in numbers will occur, followed by a slowing in the population growth rate, until some equilibrium population size is reached, the maximum allowable by the food supply. Ecologists call this number the carrying capacity. At this point, births and deaths cancel out, and the population remains the same. If population size over time is plotted for this experiment, a very characteristic curve shape, known as the logistic curve, results. Interestingly, just such a curve has emerged in modern revisions to Phillips's pioneering attempt at describing the diversity of life through time. It appears that since Cambrian times, the earth has acted like a giant jar for its entrapped species. A major difference, however, is that the earth has varied the amount of food available to its living dependents, thus changing the size of the carrying capacity. Even more entertaining, every once in a while the earth has conducted the equivalent of spraying DDT into the jar, killing off nearly everything and then letting the experiment begin anew.

John Phillips made his estimates from his own experience studying the fossil record; in 1860 there was little summary literature available about the number of species defined for any unit of time. With more than a century of fossil collection since Phillips's time, we now have far more data about species numbers and the computers necessary to assemble and sort the gigantic amount of documentation necessary to run such a study. The most recent, and by far the most thorough, examination of organismal diversity through time was conducted in the late 1970s by Dr. John Sepkoski of the University of Chicago. Sepkoski's analysis suggested that the earth's species did follow a logistic curve in the way they diversified following the inception of skeletons at the base of the Cambrian Period. Sepkoski did not attempt to study all creatures, nor did he try to base his study on species. Instead, he limited his tabulation to marine creatures, because they have the most continuous fossil record, and he studied them at the level of genera, the next taxonomic step above a species. While his curves, therefore, are only estimates of species number through time, they are powerful and probably accurate.

Sepkoski had to convince himself—and the scientific community as well—that biases in sampling have not produced a false picture of species diversity through time. Some paleontologists have argued that the apparently fewer number of fossil species found in Paleozoic rocks compared to those of the Mesozoic and Cenozoic is simply an artifact of how little Paleozoic rock there is to study. No one contests that far more younger rock still exists on the earth's surface than old rock. Hence the greater number of fossil species known from these younger rocks may simply be because more fossils have been collected from the more plentiful younger rocks. Sepkoski showed that these biases, while certainly affecting the sampling of the fossil record, are not so great as to have changed the conclusion John Phillips reached over a century ago: Through time, the number of species has continued to increase.

The curves resulting from Sepkoski's analysis suggest that the earth reached a steady state in species diversity during the Paleozoic Era and then an enormous diversity crash occurred at the end of that era—the First Event. There were, to be sure, other drops in diversity during the Paleozoic, other mass extinctions. The most notable occurred at the end of the Cambrian Period, during the Ordovician Period, and (a particularly severe one) during the Devonian Period. None of these, however, was as lethal as the great mass extinction closing out the Paleozoic Era.

Perhaps the most surprising thing about Sepkoski's study, and others that preceded it, was the finding that unlike the experiment with the closed jar filled with flies, diversity on the earth did not reach a plateau and then remain constant, or return just to previous levels after a mass extinction. Following the First Event, diversity rebounded. At first slowly, and then with great rapidity, the surviving creatures inheriting the empty, Early Mesozoic world diversified anew. By the middle part of the Triassic Period, the first time unit of the Mesozoic Era, the number of species had already risen to levels surpassing those of any period during the Paleozoic Era. And still it continued to rise. Dinosaurs began to cover the land; gymnosperm trees, the pines and their ilk, built giant forests, while marine creatures characteristic of the Paleozoic Era were replaced by a new suite of clams, snails, and coral reefs. In late Triassic times the earth was visited again by an extinction, which momentarily slowed the mad pace of diversification. But this was only a brief pause, and in the succeeding Jurassic Period the juggernaut of diversification continued anew, fueled by new types of bivalves, fish, and plankton in the sea, and dinosaurs and plants on land. The final great pause occurred at the end of the Mesozoic Era, when the Second Event obliterated a majority of species. But catastrophic as that great extinction was, it too slowed life's relentless expansion only momentarily. Following the Second Event mammals began their hegemony, in tandem with flowering plants, insects, mollusks, and an entirely new suite of oceanic plankton.

In the modern day, levels of diversity are far, far higher than at any time in the deep past. But the long history of life on the earth offers little congratulation: Many past empires of life stood tall as well, only to slide back into oblivion and dust.

6

It is the flatness of Table Mountain that you notice first. From a vantage point on the white sandy beaches to the north, you see a flat rectangular mountain rising from the sea, an angular crystal set within the confines of more rounded hills. The entire structure seems to grow out of the sea; only after coming closer can you see the low land connecting it to the African continent and the jewel of a city beneath it.

Table Mountain looms over the city of Cape Town. A cable car takes you to the top of the mountain, and from this high, windblown plateau you can see far to the north and east, into the hinterland of South Africa. As far as the eye can see rise rugged, folded mountains of sediment. They were deposited in a rapidly subsiding basin that formed about 500 million years ago; the lowest of them sit atop the granites found on the Cape Town shoreline. By starting at the base of the Table Mountain sandstones and shales and then traversing upward, into ever-higher strata, we can journey through time. We start in rocks 500 million years old, a time when the diversity curves were rising at their steepest point and the oceans were filling with species. But the history of life on land was very different; in stark contrast to the rich diversity of life in the sea, the land at that time was a barren place, where only creeping and encrusting plants eked out a fragile existence on otherwise barren rock.

Without plant life to hold it in place, any soil produced by weathering was quickly blown away in howling winds and rain; without abundant plant life to make organic humus and burrowing animal life to aerate it, the little soil that did form was almost abiotic. It was a nearly sterile land.

We have little direct information about the land during Table Mountain time, 500 million years ago. But the thick suite of quartzites and sandstones making up Table Mountain gives mute testimony to its harshness. Marine sediments accumulating next to land areas in today's world almost always give rich evidence of the abundance of life on land. The ocean's edges receive a constant supply of plant and animal life, brought out to sea by rivers and streams. These nutrients and creatures rain downward onto the ocean bottoms, there to become trapped and entrained in the marine sediments and eventually incorporated into the rock record as fossils. The Table Mountain sediments contain no such fossils. At numerous localities around the mountain are exposed great thicknesses of strata; within them are found no skeletons of land-living creatures, no logs or leaves, no nuts or insects or any remnant of land life. The traces and skeletons of marine creatures are there, but not those from the land.

The oldest sediments on Table Mountain tell us much about the world in which they accumulated. From the structures and grain size of these granular rocks, we can reconstruct a shallow seabottom, with sparkling sunlight playing over rippled sand. Standing at the base of

PHOTO BY THE AUTHOR.

Table Mountain as viewed from the south of Cape Town.

Table Mountain on a warm October day, I try to imagine my way back into that seabottom. If I take myself back into time, I may as well bring SCUBA gear; there are no excess baggage charges in imaginary time traveling.

Splashing into this Ordovician sea is a shock; here I am, 450 million years from the nearest wet suit, and the damned water is ice cold, for at this time the latitude of South Africa is nearer the South Pole than the equator. Shivering, I dive downward, ruing the lack of a companion; although many paleontologists make this trip, most stay up in the tropics, studying the grandeur of our planet's first coral reefs.

Going this far back into time is usually an anticlimax; water is water, after all, and this twenty-foot-deep seabed looks quite unexceptional at first glance. The bottom is rippled sand, with a few scattered bits of rock here and there. I dive down close to one of these and see that it is encrusted with brachiopods, small clamlike creatures that are the dominant marine invertebrates of the Paleozoic. I cluck in sympathy; the brachs will die out almost completely during the First Event, and in our world they will be only pathetic vestiges of their Paleozoic glory, living fossils eking out a pitiful existence in caves. I resume my swim above this ancient sandy bottom, noticing the wave action above me. Crossing a large patch of sand, I see a gathering of trilobites. Looking a bit like giant pill bugs, these creatures are moving slowly over the gritty bottom, ingesting sediment as they go, straining fallen organic matter from the seabottom. They are wonderful to see, for they became extinct 250 million years before the time of humans. Elsewhere the sea is empty, and that emptiness creates a nagging curiosity until I realize the reason: There are no fish here. Although nautiloid cephalopods with long, conical shells can occasionally be seen swimming awkwardly above the bottom, perhaps searching for trilobite prey, there is no flash of silver scales or darting fleet swimmers. It seems the biggest difference between my time and this ancient Paleozoic sea: Some 450 million years ago, the chordates, our ancestors, were but tiny, boneless creatures hiding from more efficient predators. In some ways it seems almost a relief; I have spent (or will spend—time traveling is confusing!) too many day and night dives in the tropics of my world trying to watch my back for the sudden movement of a shark. In some ways it is a pleasure diving in the time before fish.

7

But the time before fish didn't extend very long into the Paleozoic Era. Bits of bone begin to appear in Ordovician strata, and then whole skeletons of archaic creatures without jaws, ancestors of the detestable lampreys and hagfish, appeared in Silurian rocks, deposited 400 million years ago. These early fish looked more like clumsy dreadnoughts than the fleet swimmers of our world. They were covered with bony armor, necessary protection against myriad and hideous predators of the time such as the eurypterids, a large type of sea scorpion looking like something out of a grade-B horror movie. The great advantages of the vertebrate body plan, with its central stiffening devices, the notochord and backbone, and the superbly designed arrangement of metameric muscles allowing efficient swimming, were negated by the need for clunky body armor; the whole situation was akin to asking Carl Lewis to run a 440 wearing a suit of armor. More than 50 million years of evolution were needed to sort out the early problems of fish design. Over this great time span ever more fish species began to rely on flight in the face of danger, rather than hunkering down in armor and hoping that danger would pass. Increasing numbers of newly evolving species did so with less, rather than more, dermal bone, and increased their swimming speed and maneuverability in the process. We can still see the vestiges of the armored ancestry of the bony fish, however, in the scales and bony head regions found in modern-day fish.

It is not until the Devonian Period, the time unit following the Silurian, that fish design finally came into its own. The Devonian has been called the Age of Fishes. (That appellation could more accurately be applied to our own time, however, since far more fish species live now than ever existed in the past.) The Devonian lakes and seas were filled with an immense variety of fish; it was a great crossroads in time, for ancient lineages such as the jawless fish were commingled with new evolutionary innovations such as placoderms, the first jawed fish. Others to be found were the earliest ancestors of sharks and bony fish, the two great fish groups still present in our world. These four, separate evolutionary lineages competed for space and food, and more often than not, their favorite food sources were surely other fish. Much of this ecological battle took place in freshwater lakes rather than the sea; it seems curious that many of the fish we now associate with salt water, such as the sharks, began or underwent much of their early evolution in fresh water.

Paleontologists who time-travel back into the Devonian see wondrous things. Giant reefs composed of archaic coral flourished in warm clear water. Just off the reefs lived brachiopods and crinoids, or sea lilies; looking like giant flowers attached to long stems, these latter are related to starfish and sea urchins, and spent their lives filtering seawater to extract food. Cruising among the crinoids in search of trilobite prey were shelled nautiloids of many sizes, shapes, and shell colors, joined by their newly evolved descendants, the ammonites. Larger predators prowled these seas as well. Numerous sharks were present, as were gigantic placoderms; these creatures must have been hideous predators of monstrous appearance and size, for their largest fossils, found in dark shales in Ohio, come from fish measuring fifty feet long when alive. And darting here and there among these long-extinct beasts were fish more familiar to us, the first bony fish, whose modern-day representatives now fill the oceans and lakes and streams of our world. Most would seem quite familiar to us, if somehow brought back to life. But some would seem very peculiar, for among the Devonian faunas were fish with curious lobed fins, the stock that ultimately gave rise to the first land vertebrates.

The Devonian Period was not only a revolutionary time for life of the sea; during the same period the land, largely barren up to that time, was colonized by a series of invaders. Creeping out of the sea, advanced plants and animals consolidated gains made by more primitive creatures earlier in the Paleozoic. It is during the early Devonian Period that we find the first fossil evidence of upright, vascular plants rather than the simple encrusting forms so common earlier on. By the end of the Devonian Period, some 400 million years ago, giant trees of horsetails and club mosses rose and fell. As abundant plant life spread across the land, animals soon followed. The earliest land invaders of which we have record were scorpions, perhaps climbing ashore in search of Silurian worms and small crustaceans. But in Devonian strata are found the fossils of a new land creature. Looking otherwise much like a large fish, these rare and priceless fossils have four stout legs. On some long-ago day over 375 million years ago, a new invader crawled ashore, no doubt slowly at first, but with increasing confidence in newly evolved legs, breathing air with lungs: During the Devonian Period, the first land-living vertebrates, archaic amphibians, the ancestors of all of us, emerged from the water.

8

The highest beds of the Cape System, the thick pile of sedimentary rock beginning with the sandstones and shales at the base of Cape Town's Table Mountain, are found far inland from the sea, in a folded and contorted set of low mountains. High in these mountains a suite of strata contain marine fossils: trilobites, brachiopods, and mollusks of early Devonian age. Higher yet the strata change and the fossils characteristic of a marine setting disappear. At the time these strata were laid down, the shallow sea covering the entire coastal region of what is now South Africa slowly retreated, leaving a series of small lakes. Although sediments marking the existence of these low lakes and accompanying river systems are not rare, fossils within them are. At one place, a small patch of strata high in the mountains northeast of Cape Town, there is a thin bed of shale containing calcareous nodules. If you crack open these nodules very carefully, beautifully preserved fish skeletons will emerge. Each small fish contained in these nodules has a head and fin system not dissimilar to that of many fish living today; the tail, however, is anything but modern, having a high arcing sweep like a shark's tail. This tiny fish is known as *Cheirolepis,* and it is found in many deposits of late Devonian age the world over. It is the last fossil to be found in the Cape System of sedimentary rocks, and the last reminder of a bygone age when the most dramatic evolutionary developments were taking place underwater. Great evolutionary empires were still to rise and fall in the seas; hundreds of millions of years of evolution would pass by before the familiar marine communities of our world would eventually arise. But by the time *Cheirolepis* and its piscine contemporaries had filled the lakes and rivers of the Devonian world, the first four-legged creatures had already invaded the land and were not to be dislodged. The greatest evolutionary events of the post-Devonian world were to take place on land.

In South Africa, folded Cape System strata make up the folded mountains paralleling much of the coastline. When you finally pass through these mountains, the land flattens and becomes drier; the forests and contorted sedimentary rock of the folded mountains give way to a high plateau covering much of South Africa, a place where the next phase of evolution is written in the rocks, a place known as the Karroo.

Chapter Two

The Great Karroo

I

The Hottentots called it the Karroo, or land of thirst. Yet it is not one of the earth's most scorching depressions, a bone-dry desert like the Kalahari, for some rain does fall; nor is it an ocean of shifting sand like the Sahara. It is enigmatic and unique, this high-veldt plateau covering two-thirds of South Africa.

To most South Africans, the Karroo makes no tug on national emotions, bereft as it is of great battle sites, mineral wealth, or other claim to fame or recognition; to them it is but a large empty space dotted by lonely farms and scattered Afrikaans towns, an interval between destinations, possessing only the seemingly endless road between Cape Town and Johannesburg. It is relatively infertile, with little game at first glance. It is perhaps best known as the home of the springbok, graceful antelope symbol of a troubled nation. But to my fraternity, the world's paleontologists, the Karroo is a sacred place, for the flat sandstones and shales piled into tablelands called koppies are one of the earth's great museums, a storehouse of ancient history, holding treasures more precious than all the gold and diamonds still to be found in Africa. The earth's best record of more than 50 million years of land-animal evolution is preserved here. It is also a vale of tears, a graveyard filled with the

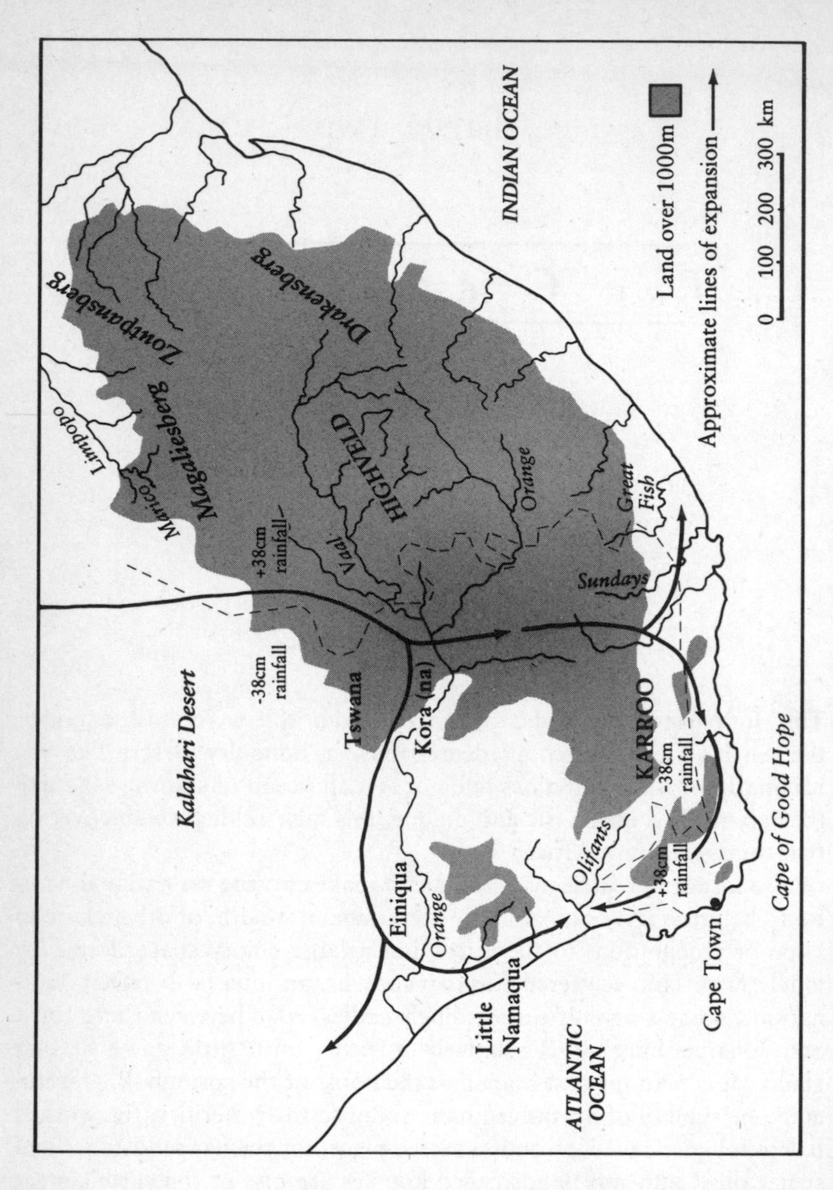

Map of South Africa, showing position of Cape Town, the Cape of Good Hope, and the Karroo desert.

PHOTO BY THE AUTHOR.

The great Karroo desert.

victims of one of the earth's great massacres: the Karroo, mausoleum of the protomammals, victims of the First Event.

2

The Karroo seems guarded by encircling mountain ranges. From the western coastline, down past the Cape of Good Hope, then running along the southern margin until once again turning northward, South Africa's coastal mountains are almost uniformly composed of lower Paleozoic sandstones and shales, sediments deposited in a shallow sea more than 400 million years ago. Cape Town's Table Mountain is one small part of this rimlike mountain chain. These ancient seabottom deposits, piled one upon another over millions of years, have been lifted from their deep resting places and now thrust upward into the sky at rakish angles. High enough to inhibit rainfall from reaching the flat interior, they have doomed the Karroo to be a nearly dry desert.

Coming inland from the sea, it takes two hours to drive through these surrounding mountains before you finally reach the Karroo itself. After the mountain passes and spectacular, contorted passages through jagged rocky vales, the land begins to settle and flatten. The treed coastline has long since changed, first to vineyards of wine grapes, then eucalyptus groves, and finally only the fynbos. This is the word the Afrikaaners use for the flora of low shrubs and bushes lining flatlands from the Cape to the Karroo; it is an incredible assemblage of endemic vegetation, for the

Proteus plants, a prominent member of the fynbos flora. In the background are seen the mountains making up the Cape Series.

fynbos flora contains as many individual plant species as can be found in the entire British Isles. During the austral spring the fynbos puts on a spectacular floral spectacle; for hours on end you pass red, white, blue, and yellow flowers of varying hue and shape. Most spectacular of all are the proteas. Looking like giant thistles invading from the psychedelic sixties, they sport every conceivable color among their large flowers. Eventually, however, even these hardy plants give way to a lower-diversity assemblage of aloe, cactus, and succulents. The sky is deepest blue and endless; the landscape becomes a vast panorama of rock and sand stretching to the far horizons.

The Karroo is many things. It is a place, and a history. But it is also an enormously thick sequence of sedimentary rock, strata that began to accumulate during the Carboniferous Period, more than 300 million years ago, and finally ceased in the Jurassic Period, nearly 100 million years later. In evolutionary terms the Karroo brackets a period of great change: When the first Karroo sediments were laid down, the most advanced land animals were squat amphibians and primitive lizardlike reptiles; by its end, advanced dinosaurs ruled the earth. The Karroo is also the place where Africa's history of continental drift is not only recorded but was first deciphered.

Because of the overall flatness of the Karroo, the oldest sediments are found at the edges of the basin. Moving into the interior the elevation rises, and you climb upward through time into ever younger rocks. Driving northeast from Cape Town on South Africa's N1 highway, you come upon the first Karroo rocks near a small village named Touws River. Along the side of the road a small, rather nondescript bank of rock marks the contact between the underlying Cape rocks and these oldest strata belonging to the Karroo. The change is abrupt and dramatic. Layered brown and tan sandstones, familiar and little different from the 150 miles of similar rock you have just driven over, around, and through since leaving Cape Town, are straddled by a rock type entirely different.

Squat black mudstones sit atop the Cape sands all along the southwestern edge of the Karroo Basin. Within these mudstones are the most amazing assemblage of pebbles, cobbles, and boulders. Such a juxtaposition of dissimilar rock types—boulders are usually not found in very fine shales such as these—can form only under very unusual circumstances. Today we find such strata accumulating at the edges of glaciers. As glaciers lugubriously crawl across the landscape, they gouge, scrape, and pick up enormous quantities of rock and gravel into their icy embrace. All of this great load is then dumped when the glacier melts. If the glacier terminates at the edge of a sea or large lake, icebergs will form, carrying with them a captive cargo of rock and debris. Sooner or later the icebergs melt, and as they do they release their lithic loads into the sea, to fall and be deposited on the seabed below. The oldest Karroo rocks were formed in similar fashion; they are the remains of a seabed that had numerous icebergs melting overhead. These rocks are sure evidence that Africa was once a great deal colder than it is today. Africa, now a continent associated with scorching deserts and humid, steaming jungles, was long ago a near-arctic wilderness crisscrossed by giant glaciers. It was undoubtedly a place of little life on land or in the sea, for the bottommost Karroo sediments are barren of fossils.

After tens of millions of years the glaciers began to lose their grip on the land, and as they did so the type of sediment accumulating in the Karroo Basin began to change as well. The seas flowed off the continent, and the Karroo became a giant landlocked depression on the southern African continent, at least a thousand miles across, and surrounded on all sides by highlands. It became a trap for enormous volumes of sediment eroding off these surrounding highlands, carried into the basin by uncounted streams and rivers.

At first this great basin was covered by saline lakes. Sediments began to accumulate in these ice-free lakes, and in them are found clear signs that life soon infiltrated into the still-frigid landscape following the retreat of the glaciers. These lake strata are thinly bedded and, from a

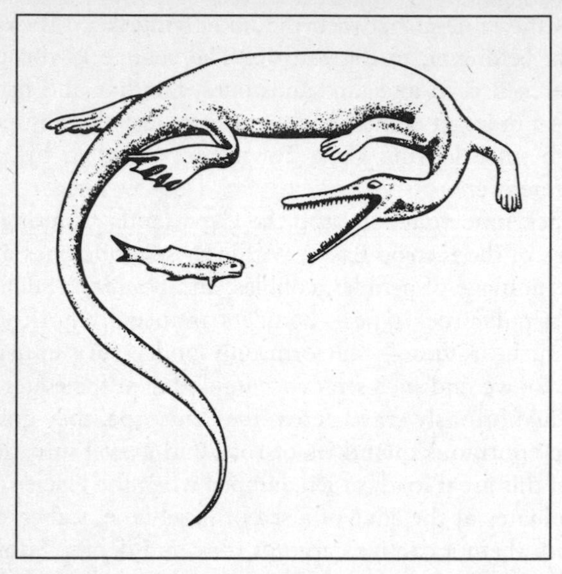

A reconstruction of the Paleozoic reptile Mesosaurus. This small creature, not more than two feet in length, became an important piece of evidence demonstrating the reality of Continental Drift.

distance, look pale tan in color. At the very top of these deposits sits a half-yard-thick bed gleaming whitely in the sun. Upon closer examination these beds are found to contain numerous interbedded ash layers. The uppermost white bed is precisely that: a thick ash layer of volcanic origin. These beds are extraordinary for a number of reasons, not the least being the role they have played in starting one of the greatest revolutions in the history of science. Over 150 years ago, a small, lizard-like fossil was found in a native village in the Karroo. The rock containing this treasure was being used as a pot lid in the village of a Griqua tribesman. The white beds were the source of this beautiful fossil. Although found in 1835, the original fossil was not described formally until 1865. Soon after this initial scientific description, an identical fossil was found in white beds lithologically indistinguishable from those of the Karroo. But this second group of white beds was found not in Africa but in South America. The fossil from both places is named *Mesosaurus;* small in size, clearly incapable of swimming the entire width of the Atlantic Ocean, there seemed no way that this tiny creature could have originated on one of the southern continents and then somehow traversed the wide South Atlantic to reach the other. Scientists of the time could only conclude that this tiny, air-breathing lake-dweller originated simultaneously in lakes of Africa and South America. But the coincidences stretched credulity, for not only were the South American and African *Mesosaurus* fossils identical, but so too were the sediments in which they were found. Astute men began searching for other explanations and, in the process, began wondering if continents could drift.

3

The idea that the continents are not fixed in place on the surface of the earth, but can somehow move about, is not new. The remarkable congruence in form between the coastlines of western Africa and eastern South America commanded geographers' attention as soon as accurate maps of the New World became available. But there have been numberless crackpot ideas throughout history, and this one seemed even more farfetched than most. The concept of continental drift first gained serious consideration in the late nineteenth century, when the then-famous Austrian geologist, Eduard Suess, suggested that Africa, Madagascar, and India were once all joined together as a single landmass and only later drifted apart. Suess based this heretical proposal on the great similarity in rock types to be found in all three areas. He named this ancient continent Gondwanaland, deriving the name from a land in India inhabited by a tribe named the Gonds.

Suess was no charlatan or crackpot, and soon a few other geologists, mainly those working in the southern hemisphere, began considering the possibility that a large continental amalgamation, also known as a supercontinent, existed in the southern hemisphere during the Late Paleozoic and Early Mesozoic eras. Australia, South America, and Antarctica were soon added to the list of Gondwanaland participants when it was found that they too showed rock structures and fossils typical of the other southern continents.

The various threads supporting the concept of an ancient, southern supercontinent were brought together in a remarkable book, published by a German meteorologist in 1912. Alfred Wegener was convinced that the great similarity in coastlines between western Africa and eastern South America went far beyond any possibility of coincidence. He amassed as much paleontological and geological information as possible to support his cause. But in many respects the crowning bit of evidence was the presence of *Mesosaurus* fossils in similarly aged beds of the various, now-separated Gondwanaland members.

Wegener had the unwavering confidence of a religious zealot. What he lacked, however, was the detailed geological knowledge necessary to truly support his hypothesis. The publication of Wegener's book was met by quiet applause and a growing sense of wonder from the southern hemisphere scientists and by deafening howls of derision from the far more numerous (and ignorant) geologists of the northern hemisphere. "How could the continents possibly move over the solid ocean floor?" cried the critics. Geophysicists, those geologists dealing with the interior and physics of the earth, were particularly damning in their criticism, forgetting that they were the same group of scientists telling the world that Darwin had to be wrong about the antiquity of the earth, which, according to their calculations, was no more than 5 million years old. (They were off by a factor of about a thousand.) When Wegener died in a balloon accident over the Greenland icecap in 1930, the geological establishment heaved a barely disguised sigh of relief. But not for long.

In the early 1900s a young South African geologist named A. L. du Toit began to crisscross South Africa, spending twenty years examining rock structure, mapping huge expanses of territory, and, in the process, filing away vast amounts of information in his encyclopedic memory. Du Toit, who soon realized that Wegener's outrageous hypothesis explained many of the geological features of southern Africa, in 1921 published his first paper about the possibility of "continental sliding." Geologists had long been puzzled about the origin of the mountains encircling the Karroo Basin; how could the Cape sandstones have been so deformed from all sides? The answer, du Toit realized, was that they could have been compressed by continental collision. He had a vision of southern Africa caught in a monstrous vise between South America and Antarctica. During the 1920s and 1930s, du Toit was able to visit other Gondwana members and see firsthand the nature of rocks of similar ages on now widely dispersed continents. Du Toit went far beyond Wegener in his understanding of Gondwanaland; he was able to reconstruct both the early merging of the various continental pieces and their climactic melding into a single continental mass in Late Paleozoic time, followed by their subsequent fragmentation during the Mesozoic and Cenozoic eras. He became versed in the stratigraphy not only of his own continent, but of the other Gondwana members as well. Perhaps his most telling argument supporting continental drift was his demonstration of the remarkable similarity in Late Paleozoic rock sequences on the various continental pieces. On each he saw a basal unit of glacial tillites, overlain by lacustrine shales containing *Mesosaurus* fossils, then deltaic and river deposits, and culminating in Mesozoic basalts. He called this the Gondwana System, known today as the Gondwana Sequence. In South Africa they call it the Karroo.

In 1937 du Toit published a monumental book on Gondwanaland, called *Our Wandering Continents: An Hypothesis of Continental Drifting*. His detractors howled him down. *Mesosaurus* could easily swim across oceans, they cried; the similarity in strata and fit of the coastlines of the various Gondwanaland members was nothing but coincidence, they scoffed. For twenty-five years the idea lay dead in the minds of all but a few diehard southern hemisphere geologists; the rest of the scientific world seems to have ignored the timely advice found on the cover page of du Toit's great book: "Africa forms the Key."

Wegener and du Toit turned out to be correct, of course. But like Vincent van Gogh, who never saw his genius acknowledged, neither Wegener nor du Toit lived long enough to see their great triumph of observation and reasoning confirmed. The proof of wandering continents did not burst into the scientific consciousness until the early 1960s, when a slew of studies brought the theory of a static earth tumbling down in disarray. First, studies on rock magnetism showed either that the geomagnetic pole had moved through time or that the continents had. Both seemed equally impossible. But in short order new evidence supporting continental drift came to light. It was demonstrated that the mid-Atlantic ridge, a poorly known line of undersea mountains running exactly down the middle of the Atlantic Ocean, was composed of a linear chain of active volcanoes constantly in the process of creating new oceanic floor. Next, a newly instituted program of deep-sea drilling demonstrated that the age of the ocean floor increased moving away from these newly discovered "spreading centers"; this discovery showed once and for all that the seafloor was spreading, and in many cases carried continents along for the ride. But where was all this new ocean floor going? Seismic studies then showed that in many places, oceanic crust dipped downward into the earth itself along long linear arcs of the earth's crust; this process invariably led to mountain chains and active volcanic mountains along these "subduction zones." Within a few short years a scientific revolution had occurred. We now know that continents indeed had drifted, and drift still. They do so because they float.

All continents are masses of relatively low-density rock embedded in a ground mass of more dense material; continents essentially float on a thin (relative to the diameter of the earth) bed of basalt. Earth scientists like to use the analogy of an onion; the thin, dry, and brittle onion skin can be thought of as ocean crust, sitting atop a concentric globe of higher density, wetter material. Continents are like thin smudges of slightly different material embedded in the onion skin. Unlike an onion, however, the earth has a radioactive core. It constantly generates great quantities of heat as the radioactive minerals, entombed deep within, break down into their various isotopic by-products, liberating heat in the process. As this heat rises toward the surface, it creates gigantic convection cells of hot, liquid rock in the mantle (a molten layer of material directly beneath the crust, which is the outermost region of the earth). Like boiling water, the viscous upper mantle rises, moves parallel to the surface of the earth for great distances (all the while losing heat), and then, much cooled, settles back down into the depths of the earth once more. These gigantic convection cells carry the thin, brittle outer layer of the earth—known as plates—along with them. Sometimes this outermost layer of crust is composed only of oceanbed; sometimes, however, one or more continents or smaller landmasses are trapped in the moving outer skin. This process, termed continental drift or plate tectonics, is one of the greatest unifying theories ever formulated by the scientific method.

Like all other continents, Africa has drifted across the surface of the globe, ambling aimlessly and randomly, a prisoner to gigantic forces of heat and convection operating far inside the earth. As du Toit suspected, in its wanderings Africa has run into other continents, and in the process its edges were crumpled.

The collision of two continents is a slow, majestic process; moving at only a few centimeters per year, thousands of lifetimes must pass before any positional change would be apparent. But as millennia pass, the continents do move relative to one another, and sometimes they collide.

The first contact of the opposing continental shelves does little. But year after year, as the two giant continental blocks of the earth's crust coalesce, enormous forces of compression act on the continental edges until the outermost regions buckle. Mountains begin to form along the two edges as the collision progresses, often creating high, spewing volcanoes amid the contorted, compressed mass of sediment and rock that was once a tranquil, flattened coastline of wide sandy beaches. Finally the two continents can be compressed no more, even though they are still driving against each other. Slowly one of the continents begins to slide over the other, often doubling the thickness of their crustal edges in the process, and then the two continents lock together, no longer able to give any more ground. A relatively recent and dramatic example is the collision of India and Asia. Forty million years ago, India was a small fugitive of the ancient Gondwana supercontinent, fleeing northward from its southern hemisphere origins until it collided with mainland Asia. In the process the edge of the Indian continent rode up onto the Asian mainland, and the result was the world's highest mountains, the Himalayas, and the thickest known continental crust on the earth.

After continents collide, one of several possibilities occurs. The newly merged continents may stay locked together and begin to travel about as one new, huge supercontinent. Or they may pull apart, splitting along the lines of the old coastlines or splitting in a different way and creating a new coastline. Such was the fate of Gondwanaland.

Knowing that Africa has changed position relative to the poles and equator greatly simplifies the interpretation of the Karroo strata. Prior to the acceptance of continental drift theory, geologists trying to explain the presence of glacial sediments in a present-day desert had to resort to much special pleading. The geological and fossil evidence makes much more sense knowing that the South Pole lay in the heart of current-day equatorial Africa for long periods during the Paleozoic Era.

The assembly of Gondwanaland—the melding of South America, India, Africa, Australia, and Antarctica into a single, gigantic continent—began during the Carboniferous Period, more than 300 million years ago. One after another, over 50 million years, the continents slowly merged together, producing great mountain ranges along their mutual coastlines in the process. Both the western and eastern coasts of Africa were compressed into mountains in this fashion, thus isolating the Karroo Basin and separating it from any contact with the sea. Sedimenta-

The White band, a prominent layer of white sedimentary rock, as seen in South Africa. The same band of rock can also be found in South America.

tion patterns changed as well. Gone were the shallow seas in which the Cape sandstones had accumulated, and gone as well were the icy glaciers producing the basal Karroo tillites. Shallow lakes and vast marshes now covered the huge basin floor. But perhaps the most significant change of all had to do with latitude. The gigantic, unwieldy continental assemblage, the now-complete Gondwanaland, itself began to move across the face of the earth as a single unit. Africa, and its newly formed Karroo Basin, began to move northward, toward life-giving warmth. And as the icy grip of the South Pole was pried from the Karroo, land life finally came to southern Africa.

4

When I was a boy, my image of prehistoric life, and especially its sequence of prominent entrances and exits, was largely derived from *The Age of Reptiles,* an expansive mural painted by Rudolph Zallinger for Yale University's Peabody Museum. Appearing in *Life* magazine and later reprinted in numerous books about prehistoric beasts, this picture (and its sequel, *The Age of Mammals*) symbolized the history of life for several generations of Americans. A truly epic painting (the original is over one hundred feet long), the mural progresses in time from right to left; things get started with an eerie green swamp devoid of animals, but you quickly climb up the evolutionary ladder as you move across the painting. Giant amphibians give way to fin-back reptiles; they in turn are replaced by a whole series of dinosaurs, culminating with a fearsome *Tyrannosaurus rex* standing in front of exploding volcanoes. Although the mural is continuous, with one fauna grading into another, viewers could nevertheless work out various intervals of past life that coincided with past geological ages. It was an icon, the definitive statement about the sequence of life in the past.

I loved this painting. It was therefore a great pleasure when I eventually saw the original. It is gigantic, covering a huge wall, in colors far more vibrant and lifelike than any reproduction could impart. Seeing it as an adult, I could easily remember myself wishing that I could go back to the various time periods depicted. Although I certainly would have liked to go back to see the dinosaurs (like all children of my time, I was a dinosaur fanatic; the current love affair with dinosaurs in no way exceeds that of my generation), my favorite part of the painting occurred early on, immediately before the Age of Dinosaurs. There, soon after the early coal-swamp forests and before the first massive dinosaurs of the Triassic Period, sits a splendid, toothy monster, named *Dimetrodon.* This great creature, with its enormous finned back, snarls at its favorite prey, a more harmless-looking but similarly fin-backed beast known as *Edaphosaurus,* while the obviously lower beasts such as amphibians can only look on from their confining swamps on stage right.

Thinking back, I can remember what so attracted me to this period in the painting. First, it looked so weird. Even in the immediately succeeding time of dinosaurs, the landscape painted by Zallinger was something recognizable, a place not too dissimilar from someplace we might find on the earth today; there are normal-looking plants, mountains, and sky. The fin-backs' landscape, on the other hand, looks like nothing on the earth today, filled as it is with strange plants and odd rock piled in purple formations. Second, the scale of the creatures involved during the fin-backs' moment in the sun seemed manageable to a little boy. Dinosaurs were just too big; my friends and I, in those days, were forever traveling back through time, and we weren't going back to study the beasts: We were going back to slay them. If I had to go back and face snarling carnivorous monsters armed only with poison-tipped arrows (the great comic book series, *Turok, Son of Stone,* was an important influence on armament), *Dimetrodon* seemed scary enough and big enough to be a worthy foe. All in all it seemed a better proposition than trying to knock off a *T. rex*.

Unknowingly, I used the Zallinger murals as my model of tetrapod, or land-animal, evolutionary progress. The amphibians first climbed out of the water (but not too far out) and gave rise to *Dimetrodon* and its cronies, who (collectively) ushered in the Age of Dinosaurs in all of its splendor. Then, for who knows what reasons, the dinosaurs disappeared. (But we all really knew—every painting and movie about the subject, including such wonderful classics as *The Animal World, The Lost World* [both versions], *Journey to the Center of the Earth, One Million B.C.* [the original, not the terrible remake], even my beloved mural, Zallinger's *Age of Reptiles,* showed some variation on the same theme: The dinosaurs were annihilated in a crescendo of exploding volcanoes and flowing lava.) Finally, after dinosaurian death, the mammals (shivering helplessly beneath the behemoths' feet during the Mesozoic), took over.

In a class about vertebrate paleontology in college, I was shocked to find out that it did not quite happen that way. A very important group of creatures is missing from Zallinger's mural. Between *Dimetrodon* and the first dinosaurs should appear some representative from the empire of the protomammals, a 50-million-year dynasty best known from the Karroo.

5

When Africa, the cold continent, slowly drifted northward more than 350 million years ago, the first land life to invade the Karroo were plants: Tough, hardy vegetation such as horsetails, mosses, ferns, and club moss slowly spread across the land. It must have been a difficult, precarious colonization, for few landscapes are more bleak or forbidding than land newly emerged from glaciation. Very little soil is present, while great piles of rock and sand litter the landscape, geomorphic refuse left by the melting glaciers. But millennium by millennium, the sun spent more time in the sky each year as the continent drifted northward; plants grew and then prospered. Soon they were not alone, for animal life followed.

The first remains of land-living vertebrates are found from deltaic or river deposits low in the Karroo succession. These fossils have been dated as mid-Permian Period in age, making them younger than the fin-backed reptiles *Dimetrodon* and *Edaphosaurus* but older than the first dinosaurs, which were not found on the earth until almost 50 million years later. In this 50-million-year period the Karroo remained a stable basin, a place of heavy rainfall and perhaps yearly floods and lush with vegetation. The land animals of the ancient Karroo Basin flourished and multiplied both in numbers and kind, becoming the richest assemblage of Late Paleozoic vertebrates known from anywhere on the earth. The Karroo is filled with their bones; at any given time, at this moment, untold millions of skeletons from that long-ago, far-away garden are baking in the summer African sun, or cracking under the harsh winter frost, eroding, disappearing to dust in the vastness of the Karroo. Like ancient starlight they pass through our world in an instant; buried a quarter billion years, they finally reach the earth's surface only to disappear in a flash of erosion. To find any given fossil and trap it from its passage back to component atoms involves nothing but fantastic chance.

Various experts have long thought that the earliest Karroo immigrants came from ancient Russia, for fossils not dissimilar to the oldest Karroo land vertebrates have been recovered there, from strata slightly older than any fossil-bearing Karroo rocks. But that view is slowly changing; sedimentological conditions in the lowest Karroo strata were not ideal for fossil preservation, and it may be that the earliest Karroo reptiles are as old or even older than the Soviet species. But that is splitting hairs; the Karroo vertebrates had to have ultimately come from someplace, since they could not have originated in South Africa. The glacial climate in the ancient Karroo Basin was too harsh, the ice too encompassing, to be a cradle of evolutionary creation. Slowly, ponderously, land vertebrates migrated into the Karroo from warmer places. They are the real missing links in vertebrate evolution, because they are found in so few and such faraway places. And they are of far greater relevance to mankind's genesis than all of the dinosaurs combined, for the protomammals of the Karroo Basin, along with the others of their kind known from various scattered localities around the earth, are rungs on our evolutionary ladder, ancestors to every mammal now alive, including humanity.

Reconstruction of the Permian finback reptile, *Dimetrodon*.

Yet for all their importance in our family tree, the protomammals rest firmly unknown and will probably stay that way. It is doubtful that anyone will ever feature a protomammal in a movie or write a best-selling novel about reconstructing one from ancient DNA. Any way you try to phrase it, the protomammals—also known as mammal-like reptiles—were either hideously ugly or faintly ridiculous-looking; they would either scare you or make you laugh, and it is my guess that very few would sell well as pets. House training and parlor tricks would have been a big problem; their brains were extremely small. And some would have been a definite menace to house guests, for many of the early Karroo fauna were meat eaters, and many were bigger than any land-living predator on earth today.

Three groups of creatures found their way into the earliest emerged Karroo landscape: amphibians, pareiasaurs (large herbivorous creatures descended from the ancestral stock of reptiles), and protomammals. The latter group received its name because of a number of mammalian features, such as having teeth differentiated for various functions. Protomammals first evolved the familiar canine teeth of many mammals (including humans), and this, more than any other feature, gives their skeletons a mammalian appearance. Other, less noticeable osteological features clearly differentiate them from the stock giving rise to the large clan of true reptiles and puts them dead center in the tree of mammalian evolution.

From the number and diversity of fossils recovered from the Karroo strata, it is clear that the three groups of terrestrial vertebrates were very numerous in the ancient basin, with protomammals most abundant of all. Most were herbivorous forms, and many were very peculiar looking. One of the most successful groups, the dicynodonts (two-tuskers) evolved a pair of large, downward-protruding tusks, while the front of their mouth evolved into a parrotlike beak; the end result looked like a saber-toothed tortoise. The tusks may have been used for digging roots and vegetation out of the ground, or may equally well have served as defense, for these mainly small, inconspicuous plant eaters were not alone in their world. Terrible predators lived there as well, veritable hellhounds from our worst nightmares: the gorgonopsids.

What an apt name! Gorgon, in ancient Greek mythology, had such an ugly head that its stare could turn things to stone. These large predators of the ancient Karroo world surely did their share of scaring, and eating.

Reconstruction of a late Permian scene in the South African part of Gondwanaland.

Large, bulky, probably not too fleet of foot if let loose in our world, they were certainly fast enough to wreak havoc and make meals among the herds of herbivorous dicynodonts.

6

On my first collecting trip to the Karroo, I came into the basin by way of a high, dirt track through the coastal mountain ranges, retracing the trail of the trekboers, the hard Afrikaans settlers of a century ago who migrated into the Karroo to escape British rule. Traveling by wagon, these proud people, mainly Huguenot descendants who originally fled Europe for South Africa to escape religious persecution, made long migrations into the South African hinterlands to find new farmlands to settle. But as the trekboers migrated inward, they came into contact with the native peoples who had long ago settled the land. The Boers began to erect fences for their cattle and sheep. With this act, armed conflict with the more nomadic Griqua, Xhosa, and eventually, the fierce Zulu tribes became inevitable. Cruel slaughters were inflicted by both sides, and neither side forgot. In many ways the ugly tyranny of apartheid is a legacy of this centuries-old conflict.

The trekboer trail brought me into the flatness of the Karroo on a fine spring day; heat had not yet come to the wide expanse of sagebrush and rock. I stopped near a large koppie, surrounded by cactus, aloe, and wildflowers. With great satisfaction I stretched cramped legs after the long drive, shucked old tennis shoes in favor of my sturdy field boots, smeared sunscreen on my arms and donned hat and sunglasses—all this ritual a pleasurable prelude to field work, the greatest joy of my adult life. I buckled on the old, polished leather of field gear smoothed by hundreds of days in hot sun or misty rain, and finally grabbed the familiar cold steel of hammer and chisel, ready now to collect, for the first time, ancient fossils from the Karroo.

Or so I thought. Searching for fossils is a matter of concentration and training. You must teach your eye to see certain signs and try to filter out the rest. Walking over the ground, you see so much that you can easily overlook the fossils. The trick is in limiting your search to several visual clues. On my first day in the Karroo, however, I hadn't the slightest idea what those clues ought to be. Sure, bones. But are they white or black?

In shale or sandstone? Crushed or whole? The group of fossils I have specialized in are ammonites, extinct marine creatures whose nearest living relative is the chambered nautilus. Their presence in sedimentary rock is usually given away by a characteristic color or shape; I scan the sedimentary rock surfaces for a shiny flash from fossil nacre or seek regular spiral shapes in the weathering shales or sandstone. Unfortunately, on my first Karroo collecting trip I could not get my mind to quit looking for ammonites. I must have passed over many bones, my mind still looking for ammonite fossils in rocks deposited far from the nearest sea; on this first day I jumped with joy at seemingly finding an ammonite (all the while my brain screaming, it can't be true, you dummy), only to find that my prize was a spirally coiled and thoroughly alive black millipede.

The strata I searched were typical of the fossiliferous Karroo, composed of alternating sandstone and shale, each formed in a different way. The sandstones are ancient riverbed deposits. Most rivers migrate across their floodplains and, in so doing, carry a deposit of sand along with them. The sandstone strata found throughout the Karroo are left behind by these ancient, migrating rivers. Sometimes the tops of the sandstone beds bear footprints, and often they contain impressions of plant stems and leaves, remains of species long extinct. Overlying these coarser deposits are dark shales, colored red, ocher, brown, or black. Originally fine silt and mud deposited during floods, these sediments have turned to hard shale over the long years. These deposits carry the most and best preserved fossil bones.

I searched the shale banks for an hour, seeing, finally, scraps of bone, but nothing worth collecting. Getting bored, I began to wonder what the ancient Karroo landscape would look like. We know a great deal about this place, for collectors have been coming to these rocks for more than 150 years to remove the skeletons of the ancient Karroo wildlife, and in the process have accumulated a huge store of information about the long-ago Karroo world. But in reality so much information must be forever lost: The colors and sounds, the many creatures without skeletons or hope of ever being immortalized by fossilization; unless a time machine is someday built, we will never have more than a tiny peek through the door of time. So I began to wonder about what I would take, given a time machine capable of going back into the past a quarter billion years. Camera, jars, recording equipment; tissue fixatives, a little food, water purifier? What the hell—I'm not finding diddly here, why not just . . .

And as I unceremoniously arrive (unlike any of the Terminators, I am at least clothed, and no blast of lightning announces my arrival, thank God), I realize that I didn't quite take all those warnings about yearly floods back here seriously enough. Rubber boots. Why didn't I bring rubber boots? And it's cold—not freezing cold, but cold enough that the shorts and T-shirt I was wearing when I left Karroo, 1991, are not going to cut it in 248,376,131 B.C.

Muttering newly learned Afrikaans profanities, I inelegantly start splashing out of the muddy swamp I have arrived in, my once-beautiful Vasque field boots now covered with black, stinking mud. The dank swamp I have fallen into is filled with decaying plant material; each step I take releases a noxious burst of methane from vegetation rotting on the bottom.

The distinctly chilly swamp stretches for another hundred yards or so, where I can just see a vague shoreline lined by short, spiky trees. I finally wade out of the swamp, happy I haven't been grabbed by some unseen, lurking amphibian. Now on firm if not completely dry ground, shivering and stinking of fetid swamp muck, I can at last look around at my surroundings. The sky is dark gray overhead, and a light, cold mist hovers in the air. I can't judge the time of day because of the cloud cover, but it seems like afternoon, and the light is similar to that of my world. My first view of animal life comes as a dragonfly, completely unremarkable, alights next to my boot. Looking around, I see others in the air, some larger than anything of my time, but in form no different. A large beetle nonchalantly strolls by, and I see a rather ominous-looking centipede nearby. Other than that, zip. There are no birds, no snakes or turtles visible, no frogs. I soon learn, however, that although this seemingly perfect frog pond is frogless, it is not without other amphibian charmers.

I almost jump out of my muddy boots at the sound of a startlingly loud thrumming, coming from a pool several yards to my left. There, in that rather big puddle, I see a large green salamanderlike creature, but of a size unlike any salamander I have ever seen—the damn thing is four feet long. It is some sort of labyrinthodont amphibian, something akin to *Eryops,* I suppose. Several others of its kind are nearby, tympanic membranes pulsing, bulbous yellow eyes bulging. No warts, though. I involuntarily jump again as the nearer amphibian opens its mouth, flashing a huge number of wicked, spiky teeth in the process, and lets out another bass-drum bellow. This produces much consternation among the assembled amphibians lining the pool. I assume that the mating season is in full progress.

Climbing toward dryer land, I set out to find the more terrestrial members of this fauna. The landscape is a riot of vegetation, but much of it is low creeping plants and bushes, with many spiky ferns and horsetails flourishing in the wetter ground. Larger trees occur in clumps, and there is a variety of archaic-looking forms: Seed ferns seem to be present in abundance, and many trees look like cycads. In the wet ground around the trees are numerous burrow openings, but I am not too keen on sticking my hand into any of them to arouse whatever inhabitant may be domiciled within. So I keep walking, the edge of the swamp in sight, and soon come across my first protomammals.

They are dicynodonts, judging from the impressive pair of tusks curving down from the upper palate. Low-slung animals, they exhibit a waddling gait as they slowly move over the landscape, browsing among the low vegetation all the while. About two feet long, these small creatures seem none too concerned about my arrival in their midst; they ogle me incuriously from time to time with their bulbous, protruding eyes, then return to grazing. They are clearly part of a herd, and I count several dozen, with more farther off. A much larger pair of dicynodonts of a second species is present in the distance, but these too seem to be grazers. All in all it seems a cold, wet, stolid yet admirably peaceful assemblage of animals.

The dicynodonts don't look reptilian—in fact, they don't look like anything at all of my world. They are covered with short fur and are a dull tan color. At first I think that they do indeed look more like mammals than reptiles. But then I move closer to one and see that other than the large pair of tusks, the creature has no teeth; the snout is composed of a hard, beaklike material, more like the bill of a parrot than the mouth of a mammal; the duck-billed platypus is somewhat similar. I cannot detect any sexual dimorphism, and there seem to be very few juveniles among the herd. Most graze on low shrubs bearing long, oval leaves; I guess that these plants are *Glossopteris,* a plant common to and characteristic of the Gondwana continents.

I am just about to photograph the nearest of the small dicynodonts when a hideous, throaty roar erupts from somewhere nearby, followed by the noise of a fearsome thrashing among the vegetation. At the first sound of this bestial scream the dicynodonts around me begin to scatter, bleating pitifully as they scuttle on their stumpy legs in all directions. I am now ruing the lack of an AK-47, because I figure This Is It, my first introduction to ancient predation, with some monstrous gorgonopsid or large dicynodont predator about to burst on the scene, warm scent of a *true* mammal spurring it on. But after my initial fright, nothing appears, although the fearful roaring and thrashing continues unabated, seemingly coming from just behind a glade of bushy lycopsid trees nearby. Creeping forward to what I am sure will be the scene of some predatory attack by one of the big carnivores of this world, I manage to peer around the spiky copse of archaic, segmented trees. I am transfixed by the scene: Not ten yards in front of me sprawls a four-foot-long gorgonopsid, one of the top predators of this world, roaring blue murder from its hideously contorted, fanged mouth; it is squashed onto its stomach, all four legs splayed out, with a second large creature firmly astride its back. This second figure resolves into a human torso, clad in denim jeans and workshirt. Damn! I thought I was the first scientist back here. Sitting atop this beast, facing its tail, struggles a very large man, straw hat on his long-haired head, beard rippling in the cold breeze. The big man looks like a cowboy riding backward on a bucking bronco, his weight barely sufficient to keep the ugly predator pinned to the ground. I then notice that the man has a long thermometer in his hand and is trying to take the beast's temperature. There is only one scientist so worried about the body temperatures of ancient creatures—it must be Bob Bakker!

7

In the proud tradition of American dinosaur hunters he is larger than life, filling up lecture halls, books, and controversies with an inescapable presence; there is no small way to describe or write about Robert Bakker.

I first met Bob Bakker at some symposium or conference in the late 1970s; in a large room among a horde of networking, drinking, gossiping paleontologists he held center stage. I walked up to this man, whose early papers I had so admired, and asked the first question coming to mind: "What killed off the dinosaurs, anyway?" I was astounded when he stared at me fiercely, called me an idiot, and stalked off. My first impression of Robert Bakker was of amazingly long hair and very short patience. I was somewhat mollified when a fellow grad student told me that he treated everybody like that.

I got a chance to ask the same question a second time, ten years later, when Bob Bakker came to my university to give a lecture about dinosaurs. This time around he was much more mellow, if not any less endowed in things tonsorial. The lecture was fantastic, the auditorium packed, and the audience clearly disappointed when Bakker finally finished: They wanted more. Bob Bakker stayed several days with us, giving me the opportunity to rephrase my original question diplomatically (I got a long, interesting answer this time) as well as ask many others, learning, in the process, a great deal about dinosaurs, their world, and current research being conducted by the scientists studying them. After his visit I found that much of this information and more is presented in his wonderful book, *The Dinosaur Heresies,* a rare blend of revolutionary science presented in entertaining form. In many ways it was because of this book that I decided to visit South Africa and its fabulous Karroo fossil deposits.

Robert Bakker is a student of the dinosaurs. While the Karroo does indeed bear the bones of dinosaurs amid its ancient, eroding sediment, they are among the youngest and last creatures to have inhabited the great basin. But it was not Karroo dinosaurs that brought Bakker all the way to South Africa. He came to study fossils of creatures living there long before the first dinosaurs—the protomammals. Much about the protomammals links them to their descendants, the true mammals; many characteristics of osteology and structure are quite similar. Bakker, however, searched for a link much more tenuous, a characteristic that, at first glance, would seem about the last thing that could ever be preserved in a bony skeleton: Bakker searched for evidence showing that the protomammals of the Karroo were the first creatures on the earth to be warm-blooded.

Bakker's great gift is that he has approached his subject—mainly dinosaurs, but other prehistoric vertebrates as well, including the protomammals—from a variety of viewpoints. His method in proving that neither dinosaurs nor protomammals were cold-blooded, like all members of today's reptilian fauna, was to attack orthodoxy on a variety of fronts. Warm-blooded creatures, such as mammals and birds, maintain a nearly constant body temperature in any range of heat or cold. This trait sets them apart from most other forms of life, which have body temperatures equal to that of their surrounding environments. Dinosaurs and protomammals have long been assigned to the Class Reptilia, and since all currently living reptiles, such as lizards, turtles, and snakes, are cold-blooded, it has been assumed that extinct members of the class had similar metabolisms. During the 1960s and 1970s, however, a number of scientists studying dinosaurs began to wonder if this long-held view was actually true. In our world, very few reptiles are found in frigid climates; cold-blooded land creatures don't work very well at low temperatures. During the Mesozoic, however, dinosaurs seemed to have been distributed just about everywhere on the earth—including at very high latitudes. Even though we know that the earth was warmer during the Mesozoic Era than it is today, the arctic regions, with their many months of twenty-four-hour darkness each winter, would still have been cold and forbidding places for a cold-blooded dinosaur. Yet dinosaur fossils have been found from strata located north of the Arctic Circle.

Over the last two decades many different workers have explored the possibility of warm-blooded dinosaurs. None, however, has approached the problem from so many lines of evidence, or with such zeal, as Robert Bakker. His hypothesis—that dinosaurs, like mammals and birds, could maintain a virtually constant body temperature regardless of ambient temperature—rested on four lines of argument that can be tested using fossils:

1. The bone structure and internal anatomy of warm-blooded animals is different from that of cold-blooded ones.
2. Warm-blooded creatures grow faster than cold-blooded forms.
3. Predator-prey ratios are different among warm-blooded and cold-blooded food chains.
4. Warm-blooded and cold-blooded animals show different evolutionary rates and susceptibility to extinction.

Being warm-blooded is a tremendous ecological advantage. In the morning, a cold-blooded creature, especially after a cold night, must somehow raise its body temperature high enough to get going. Lizards are terrible morning creatures: They must first sun themselves for some time before they can set out in search of breakfast. And if you like lizard soup or lizard fricassee or just plain raw lizard, morning (as well as night) is a great time to be a predator. Mammals, on the other hand, are capable of rapid movement and activity even after—or during—the coldest of nights. Mammals, however, have a terrible metabolic price to pay: They require enormous amounts of energy to keep their metabolic fires constantly stoked. A very large percentage of the food that we and other warm-blooded animals eat is used simply to keep us warm. Consequently, all warm-blooded creatures must eat more (about ten times more) and more often than a cold-blooded being of the same body weight. And not only do we require more food: The actual maintenance of all this body heat necessitates a very different type of metabolism—a more efficient and hotter furnace, if you will. While no warm-blooded creature has a special body organ responsible for maintaining all this extra heat ("I'm sorry, Mrs. McNab, but your husband's pilot light went out during the night, and now he's cold-blooded"), we can see physical differences in bone structure between warm- and cold-blooded animals that are also detectable in fossils. These differences thus serve as an important tool in deciding which metabolism was characteristic of various creatures in the past. Bakker and a French scientist, Dr. Armand de Ricqlès, examined bone from a wide variety of modern and fossil reptiles and mammals, and found that dinosaurs had bone structure typical of warm-blooded creatures. In later work, they found that the Karroo protomammals also had bone structure far more typical of warm-bloodedness than cold.

Warm-blooded bone structure, and other anatomical features typical of mammals, such as the presence of a diaphragm and a secondary palate allowing simultaneous eating and breathing, are found in both Karroo protomammals and true mammals. Zoologists today will place an animal in the Class Mammalia only if the species in question both suckles its young with milk and has a very particular arrangement of bones in the middle ear. But paleontologists have yet to find their first fossil breast, and since Permian Karroo protomammals did not have the requisite ear bone structure—which first appeared among their Triassic descendants—they cannot be classified as true mammals. Yet Bakker and others suspect that the protomammals were very close to the mammalian grade of organization—and were certainly warm-blooded. They drew this conclusion from two lines of evidence. First, analysis of bone structure strongly suggests that the protomammals were warm-blooded. But perhaps an even stronger case for this, according to Bakker, comes not from anatomy but from a very different source: the ratio of predators to prey.

In any ecosystem there is a flow of energy, where sunlight and nutrients are transformed into living tissue by photosynthetic plants. This plant tissue, the base of terrestrial and marine food chains, can then be consumed by herbivores, which are in turn eaten by carnivores. Each of these feeding types is part of what is called a trophic level. In a perfect world, each of these steps might involve a complete transferral of energy, where every pound of plant is turned into a new pound of herbivore, and every herbivore is transformed without loss into carnivore. In reality, there is nowhere near such efficiency. As a rule of thumb, ecologists suggest that the efficiency of energy transferral at each trophic level is about 10 percent, so that one hundred pounds of plant is turned into ten pounds of herbivore, which in turn becomes one pound of carnivore. In nature, however, the transfer of mass and energy from system to system varies widely and depends to a large extent on the individual metabolisms involved. Among vertebrate assemblages, for instance, being warm-blooded or cold-blooded is a very important factor in determining the relative biomass of various trophic levels. Ecologists have studied mammalian ecosystems at length. One of the most well-known comes from the plains of central Africa, where abundant grass and vegetation support a diverse and populous group of mammals. If you add up the weights (biomass) of the various herbivores (elephants, giraffes, buffaloes, antelope, deer, etc.) and then the weight of the carnivores (lions, leopards), you find that the carnivores make up only about 1 to 3 percent of the herbivores' biomass. In energetic terms, a lion needs so much energy to keep its body warm that it has to eat a great deal of meat: According to Bakker, it takes 20,000 pounds of meat to keep 1,000 pounds of lion fed each year. Although there is certainly a great deal of game on these African grasslands, their numbers are not limitless. Consequently, there are not very many lions; the ecosystem cannot support many carnivores as voracious as a lion.

Cold-blooded ecosystems are different. Because cold-blooded carnivores eat relatively less than warm-blooded ones of equal weight, a given prey population can support far more of them. To test his ideas about predator-prey ratios, Bakker searched for examples of ecosystems where cold-blooded animals are the top carnivores. He settled on an ecosystem with spiders and insects, finding that spiders can make up as much as 20 percent of the entire arthropod fauna. (This example finally made it clear to me why my house seems overrun with spiders even in the apparent absence of any visible food for them.)

Another such ecosystem, however, may be of greater relevance to this test. In 1986 and 1987 I took two delightful research trips to a place called Lizard Island, on Australia's Great Barrier Reef, to conduct research on deep marine mollusks found along the reef edge. This small, very appropriately named isle, no more than five miles across, is many miles from the mainland and, except for the visiting scientists and tourists, has no mammalian creatures whatsoever. What it *does* have is a diverse and abundant lizard population. They were everywhere. I lived in a small hut, by the beach, filled with lizards: Most were small geckos and were welcome guests, for they ate small insects and chirped happily at night—until, one fine evening, a fat green specimen fell off the ceiling, landing in the middle of a newly arrived *pièce de résistance* just put on the dinner table, a fine seafood curry. From that point on the geckos were *lizarda non grata*.

Other lizards, however, also made journeys into my hut. The largest on the island were varanids, also known as monitors, most of very impressive size: Four-footers were common, and larger specimens sometimes could be seen. These large species are near relatives to the world's largest living lizard, the Komodo dragon, which is found only on several small islands in Indonesia. The Komodo dragons are very impressive because of their large size (they can attain the size of a small horse), and are rather repulsive beasts as well: According to published reports, they stink hideously, are constantly extruding a large forked tongue, and will carry off small children if given the chance. They are ambush hunters: Lying in wait until some small animal walks by, they will rush and kill their prey, using an impressive mouthful of teeth. They carry the prey off to some hiding place, where they eat as much of the poor victim as possible. Then the lizards fall asleep amid the rotting carcass, awaking occasionally to eat a little more aged meat.

The Australian version of these monitors would come into my house occasionally, always at night, and in the process scare the bejesus out of me. I was awakened in the night by a quiet slithering and scratching, sounds made by the clawed feet and dragging scaled tail of a monitor stealthily crawling under my bed, looking for reptilian fare, no doubt constantly flicking its forked tongue in the darkness. Naturally enough I began to (carefully!) watch these lizards, noting that they often could be seen eating smaller cousins, siblings, and perhaps progeny. I sat on a beach once, watching the movements of an engaging, small brown lizard on the sand, only to see it wolfed down by a huge monitor that had apparently been lying in wait in nearby bushes. The monitor came out of the vegetation in a rush, grabbed the small lizard by the midsection, and then began to swallow it headfirst, finally walking back into its bushy lair with a long, feebly twitching tail still hanging out of its mouth.

What impressed me most, during my very unscientific survey, was how many large monitors there were relative to the number of smaller lizards visible. Perhaps I just noticed larger lizards more (a natural enough reaction toward creatures crawling around under your bed at night), but my qualitative observations are borne out by the study of other food chains involving lizards and are exactly Bob Bakker's point about why dinosaurs and most protomammals could not have been cold-blooded: Cold-blooded predators have such low metabolisms that they do not eat often, and thus a given prey population can support many more of them than it could of warm-blooded predators. If dinosaurs or protomammals had been cold-blooded, we should find the fossils of many more predatory types of both than have actually been recovered.

At what point did the transition from cold blood to warm take place? If the Karroo protomammals were warm-blooded, when did this critical transition happen?

The earliest four-legged creature climbing out of some primeval Devonian swamp, about 400 million years ago, was an amphibian, and certainly cold-blooded. But so successful was this creature and its novel innovation, the adaptation for life on the land, that it quickly spread across the world, spawning many new species of four-legged vertebrates in the process. All amphibians, however, require a moist habitat and water to breed in. The earliest land colonists could never stray far from the swamps and lakes of their ancestry. Two innovations were necessary finally to free the early land vertebrates from the shackles imposed by their need for water: a tough, drought-resistant skin and an egg capable of developing on dry land. Both of these adaptations were evolved by late in the Devonian Period or early in the succeeding Carboniferous Period (the Coal Age), perhaps 375 million years ago. The creatures that first evolved these two liberating traits were the earliest reptiles.

Evolution within reptile and amphibian lineages proceeded rapidly during the Carboniferous Period of 360 to about 290 million years ago; the largest animals appearing during this time were the fin-backs, with my beloved *Dimetrodon* clearly King of the Jungle and Swamp, the top carnivore of the land. *Dimetrodon* had a large, doglike head and is the direct ancestor of the Karroo protomammals—and thus is an ancestor to all mammals now living. But in structure, anatomy, and Bakker's measure of predator-prey ratios, all the amphibians and fin-backs were surely cold-blooded. Museum collections of fossils from this time show that numerous predators coexisted with the plant eaters—as many as 25 percent, according to Bakker's studies. This indicates a food chain capped by predators having a cold-blooded metabolism. If Bakker's hypothesis about the connection between cold- or warm-bloodedness and predator-prey ratios is correct, it means that the first 100 million years of land vertebrate history was entirely cold-blooded.

The world's best-known fossil record of this long, first interval of land life comes from the deserts of western Texas, where dark sandstones and shales have yielded a spectacular treasure of skeletons from the days of *Dimetrodon*. But younger sedimentary rocks in this area contain few fossils; the conditions that had produced a thick stratal accumulation packed with reptilian and amphibian bones must have changed. To study the next phase of vertebrate evolution one must travel either to Russia, or better yet, to the Karroo.

Three aspects of the Karroo fauna differentiate it from the earlier fin-back assemblages. First, there were so many more species among the Karroo protomammals than there were during *Dimetrodon*'s days. Second, the Karroo protomammals showed very high rates of evolution (many new species were formed over relatively short periods of time) and high rates of extinction; together these two attributes meant that there was a large number of short-lived species. Finally, there were relatively few predators among the protomammals—Bakker's studies of museum collections in South Africa showed the figure to be about 5 to 12 percent by number or weight, compared to the 25 percent in the fin-backs' world. Somewhere, between the time of *Dimetrodon* and the beginning of the protomammals, a revolutionary innovation took place: the evolution of warm-bloodedness. Perhaps it happened on the long march into the Karroo by the first immigrants, or perhaps it was an adaptation to the cool climate of the Karroo Basin. The end result was the same: The attainment of warm-bloodedness opened the doors to formation of many new species among the protomammals. But this fantastic innovation has seemingly carried with it a deeply buried but monstrously evil side effect, a curse still rolling down through the ages: Warm-blooded faunas are far more susceptible to mass extinction than are their cold-blooded ancestors. Warm-bloodedness let vertebrate animals conquer the earth. But the warm-blooded metronome of evolution sometimes beats out a cadence of death, decay, and mass extinction.

8

Soon after arriving in the Karroo Basin, the protomammals began to proliferate both in number and kind. The earliest arrivals were faced with a giant, empty, warming land, filled only with growing and multiplying plants. New opportunities for food stoked the evolutionary flames, with warm-bloodedness the fire. Over unnumbered generations they spread across the thousand-mile-wide basin, small populations getting cut off from ancestral herds, reacting to new environmental challenges, and evolving to overcome them. New species evolved at a rate not seen on the earth since the heady days of the basal Cambrian explosion, when skeletonized life filled the seas through a rush of speciation. So too did the protomammals rush to new forms.

The placid protomammal herbivores of the Karroo were not the only arrivals into this promising land. Fierce hunters came with them and, once in the basin, began their own evolutionary dance. An arms race began, where evolutionary move was checked by countermove; new, more wicked teeth were countered by greater fleetness of foot or better camouflage; better eyes and sharper claws brought about herd behavior or adaptations for burrowing. As new, more efficient predators and prey evolved through the process of speciation, the less well adapted forms often fell into the pit of extinction or survived by adopting a new habitat or behavior. The protomammals' pace of evolution moved to a quickened beat, far more allegro than the long Coal Age dirge of their cold-blooded, reptilian ancestors; they raced through time, many species lasting but a few million years before a swifter, smarter, better-adapted model drove the old into the junk heap of evolutionary obsolescence. And so this merry dance progressed, predator and prey, for perhaps 15 or 20 million years, a constant progress toward increased diversity and efficiency; in a word: modernization. And then, for the first time in this world, disaster struck.

In the middle part of the Permian Period, about 250 million years ago, and perhaps 20 million years after the first protomammal emigrants reached South Africa, an episode of extinction occurred among the most diverse assemblage of protomammals ever to inhabit the earth—perhaps one hundred different species in the Karroo alone—and a variety of cold-blooded amphibian and reptilian cousins. Many species disappeared rather suddenly from the Karroo's fossil record, quickly replaced in overlying strata by new species. Though not a truly catastrophic event, for this extinction pulse did not affect even half of the species, it was nevertheless unprecedented: For 100 million years, reptilian history had been an unbroken record of diversification. Things returned somewhat to normal, if species diversity was somewhat lower than during the mid-Permian heyday, for perhaps 5 million years. And then the darkness of extinction rolled across the land once more, but this time in deadly earnest; species after species fell away, not to be replaced for a very long time. Death stalked the land, reaping a grim harvest among hunter and hunted alike, protomammal predators and prey dying, rotting, scavenged; becoming bleached and separated bony clasts disintegrating where the animal fell, or carried by the winding rivers or yearly floodwaters to sandy or muddy graves, there to rest for 245 million years, finally to arrive in our world as a muted echo of the long-ago First Event. During that time of death the future of humanity was in the balance and nearly ended, long before it ever had a chance to begin. But in that great carnage one protomammal carnivore and one larger herbivore escaped the extinction's cruel grasp. One was our ancestor, the other its food. Life all over the earth was snuffed out. And we don't know why.

9

What-ifs are among the most futile yet common forms of human exercise. Nevertheless: What if the great mass extinction smiting the protomammals and so much else 245 million years ago had not occurred? Would the history of mammals have played out in the same way, only much sooner? It took 65 million years of evolution, following the death of the dinosaurs at the end of the Mesozoic Era, to produce the full suite of mammalian species currently on the earth. Might there have been elephants and lions and shrews and rats, whales, bats, squirrels—and humans—180 million years ago? Was the evolution of intelligence on this planet delayed by almost 200 million years because of the great mass extinction ending the Paleozoic Era? Or was the first true mammal, whose skeleton we find 10 million years after the First Event, in Karroo sediments over 200 million years old, in some way a product of that first great mass extinction? Was the evolution of parental care and efficient thermoregulation and all other features we consider hallmarks of "mammalness" forged in the fires of desperate survival during competition with the emergent dinosaurs, themselves descended from a single survivor of the First Event?

Stephen Jay Gould, the great Harvard evolutionist, maintains that evolution is a chancy, nondeterministic business and that it would never play out the same way twice, if there was some way to reset the record. I have no doubt that Gould is correct. But chance is a funny thing; someone wins a million-dollar lottery every day. What could we have done with that 200 million years had the First Event not happened, and had we (or something else) evolved intelligence during the Jurassic Period? To our knowledge, no species on the earth has ever survived 200 million years; the average longevity of a mammalian species is less than 5 million years. But could that be our path to immortality, not as individuals but as a species? In that time would humanity have climbed upward from this planet to colonize the distant stars? Or, like the actual inheritors of the earth following the First Event, would our history have ended 65 million years ago, as the dinosaurs' did, by a hammerblow from space?

Chapter Three

End of an Era

I

Vermilion shadows slowly crawl across the Karroo desert floor as the afternoon ends. Far below me the expansive valley begins to change color and seemingly gathers new texture in its weave of sage, cactus, and aloe. The layered strata of the far hills also take on a new hue as reds and purples begin to color the scree-covered slopes.

In the failing light it is pointless to continue my search for fossils, so I put down gathered treasures and wearily take a seat on a convenient sandstone stool, looking out over the Karroo. One thin road snakes through the valley, and far in the distance sits a lone farm. With its oasis of lush green trees, it seems an incongruous emerald set amid the dry parchment of desert.

I have spent the day searching for bones, stony bones of great age buried in the rocks of the Karroo. It has been a wonderful day, and a troubling one. On a perfect day to collect fossils, with high fleecy clouds covering the sun whenever it threatened to become hot, my companions and I sampled various stratigraphic horizons deposited in the few millions of years prior to the First Event. The fossils have not been rare; I found two skeletons of tiny, burrowing protomammals, their delicate, white bones enclosed in dark silty matrix. And we discovered bigger fossils as well: On a wide expanse of flat sediment, we found the bones of a creature much larger, the remains of a hulking dicynodont predator as large as a lion. Its white gleaming skull glared out of the rock; like a spectral being on Halloween night it seemed to crawl slowly out of its quarter-billion-year-old grave, but vainly: This creature is not to be reconstituted. Its bone is harder than the surrounding matrix, and thus has been slowly revealed as the passage of time stripped away surrounding cover, but the agents of erosion—the wind, rain, heat, and cold, the action of groundwater and chemical destruction—are more than sufficient to destroy this priceless skeleton. We had no tools to excavate it, nor the three days it would take to do so if we did; this large skeleton deserves to live ever onward in a museum but instead will return to wind-borne dust. The Karroo is packed with spectacular fossils, definitely nonrenewable resources, silently calling out to the few paleontologists who wander this land, emerging for a brief instant before crumbling away. But it is not the loss of these skeletons that is disturbing; the Karroo strata are thick, and packed with bones, and there will be enough to collect for many millions of years, if mankind or any future sort of gravediggers so choose. I am more disturbed at the seeming rapidity of the event that struck down the protomammals in the long-ago Karroo and the implications this holds for our world.

We spent part of the day looking at black shales overlain by coarse brown sandstones innocuous enough, seemingly no different from rocks below or above. First in a high mountain pass and then along an angular koppie wall; in a roadside pit and then visible in branching streambeds we saw the same succession of rock, and the same grim message: A diverse assemblage of protomammals below and but one common fossil species above, this passage marks the boundary between the Paleozoic and Mesozoic eras.

A diverse and populous fauna made up the last assemblage of Paleozoic land vertebrates. Composed of grazers large and small, with burrowers underfoot and predators in their midst, the many species and individuals inhabiting the Karroo some 245 million years ago were brought down by the First Event. Those scientists who have studied it universally believe that this great extinction took place over a long period of time; paleontologists have been perplexed by the savage intensity of the First Event in the absence of clearly identifiable cause, but have taken comfort in its great duration: A period of death lasting as much as 10 million years seems less frightening than a sudden killing. But everything I have seen today argues a quick, savage murder, not a death in a long, peaceful sleep.

After visiting the Karroo and looking at the stony layers recording this greatest catastrophe in the history of life, I wonder what rocks my predecessors have been looking at.

2

My first trip to the Karroo was geological tourism. On this, my second trip, I have a definite agenda. I am here to look at the geological beds marking this great extinction with a specific question to answer: How long did the extinctions among the Permian protomammals take?

Elsewhere in the world, the First Event has been studied for well over a century. It has proven to be a frustrating job. During those decades of work, as region by geological region containing rocks of Late Paleozoic and Early Triassic age were collected, mapped, and described, it became evident that no sequences of marine sedimentary rocks had been found that had been deposited at the height of the extinction. In every case, the most critical interval of time was unrepresented by rock.

The geological record is by no means complete; sedimentary rocks have not accumulated every day of the earth's history, and much rock recording important epochs has been destroyed by erosion, mountain building, or metamorphism. But more often than not, given enough searching, a stratigraphic section containing a previously missing or critical age is found somewhere eventually. Yet that key section of latest Permian and earliest Triassic age—marine rocks deposited at the height of the First Event—has proven maddeningly elusive.

By the middle part of this century it was clear that all Late Paleozoic strata in the United States and Europe contained the missing interval. The strata of youngest Paleozoic age that are preserved on these two continents contain fossils typical of the era: brachiopods, crinoids, trilobites, followed by a sharp break. The overlying strata are of a different rock type and contain a different fauna—one typical of the Mesozoic Era, characterized by clams, ammonites, echinoids, and new types of fish. The change in lithology and fossils is a clear indication that the critical interval of sedimentary rocks is missing. Either it was never deposited, or it was deposited but subsequently removed by erosion, leaving a gap of 2 to 5 million years. What happened during that critical time in earth history? Did the Paleozoic creatures slowly die away over the many years, one after another, to be replaced gradually by new species that we now associate with the Mesozoic Era? Or was the Paleozoic fauna destroyed by a catastrophic event of much shorter duration? This most critical question, if ever we are to unmask the killer, will be answered only if the missing interval in time and strata is located somewhere.

In frustration, geologists began traveling farther afield, searching for a complete section of Permian marine sedimentary rocks. The search took investigators to exotic and rugged locales: Greenland, Pakistan, Iran, and Kashmir were visited and studied, and in each case the strata were found to be incomplete. But in the late 1970s, a breakthrough seemed at hand: Extensive and thick marine sedimentary rock covering large areas of China was discovered to be of Permian and Triassic age, and when first studied, these beds appeared to fill the missing gap. There was great excitement when Chinese geologists found, and described, a thin, white clay band sitting atop the highest of these beds; there was greater excitement still when they announced that enhanced concentrations of iridium and platinum had been discovered in this layer, for geologists in the early 1980s had just discovered that similar clay layers were associated with another great extinction: the Cretaceous-Tertiary extinction, the Second Event.

In 1980 Luis and Walter Alvarez and colleagues from the University of California at Berkeley had startled the scientific world with their theory that the great extinction closing out the Mesozoic Era had been brought about by lethal effects following the impact of a giant meteor with the earth. They came to this startling conclusion after finding concentrations of platinum and a similar metal called iridium (both rare on the earth, but common in asteroids) in an Italian clay layer of Late Cretaceous age. These elements were found in the stratigraphic bed exactly coinciding with the mass extinction. Based on this single discovery, the Alvarez group made a bold pronouncement: They predicted that similar clay layers of exactly the same age would be found elsewhere on the earth, all created as meteor and crater debris fell out of the stratosphere following the explosive impact of a meteor, calculated to have been at least six miles in diameter. So catastrophic were the effects of this impact, said these researchers, that over 50 percent of species on the earth, including all dinosaurs, became extinct. When similar, iridium-bearing clay layers were in fact found atop Cretaceous rocks around the globe, a unifying theory of mass extinction seemed to be emerging: Mass extinctions were caused by asteroid impacts, leaving in their wake thin clay layers packed with platinum and iridium, and lots of dead bodies. Thus, the Chinese announcement in 1982 that they had found a similar clay layer dating from the Late Paleozoic, also packed with extraterrestrial material, seemed to confirm that the First as well as the Second Event had been caused by meteor impacts, one hitting 245 million years ago, the second 65 million years ago. But subsequent analyses of the Chinese clay layer showed the first ones to have been in error; there were no traces of extraterrestrial material, no fingerprints left by a comet or meteoric impact. And further study even showed that the Chinese sections were not complete, as originally reported, but like all others contained a missing interval at the top of the Permian. It was as if some malicious Supreme Being had capriciously ceased all marine sedimentation at the height of the great death that ended the Paleozoic Era, removing any record of the critical interval that could answer the most pressing question of all: Why did it happen?

Until we have some idea about the pace of the extinctions, the time span over which mass death ravaged the land and sea, we cannot answer why. At the present time all we can do is make a body count. The list of deaths include, in the sea: all corals and trilobites; virtually all sponges, bryozoa, brachiopods, crinoids, echinoids, ammonites, foraminiferans, ostracods; and a majority of fish, snails, and clams. In the early 1970s, University of Chicago paleontologist David Raup made the most comprehensive estimate of species death during this extinction. He concluded that over 90 percent of marine creatures died out.

After the First Event, life returned to the seas very slowly. It took 10 million years before coral reefs once again reappeared and even longer to restock the benthos to levels found prior to the extinction. Not only was diversity greatly affected, but the composition of creatures as well, for so great were the marine extinctions at the end of the Permian Period that they completely reset the composition of subsequent marine life. The Mesozoic oceans contained a suite of creatures almost completely different from those of the Paleozoic.

And what of the land? Were terrestrial sediments deposited without interruption during the critical interval of time, the acme of extinction seemingly unrecorded anywhere in marine sediment? Currently, great controversy brews about the severity of the extinctions on land; so great is our ignorance of the event that we cannot, as yet, even decide if the great mass extinction affecting the seas took place at the same time as the extinctions on land. We know only that virtually all protomammals died out either at the very end of the Permian Period or soon before. But how soon before? That question can be answered only by slow, painstaking geological research of the critical time interval preserved in strata, patiently excavating, bone by bone, the gravesites of potential witnesses to the event. To our knowledge, the best place on the earth to conduct such research is in the thick strata making up the Karroo. I was amazed and dismayed, upon finally reaching South Africa, to find that even after 150 years of paleontological research among the Karroo's ancient graves, this work had not yet taken place.

3

The noise is jarring, a high-pitched, visceral cacophony. It sounds as if some gigantic insect is just outside the door, trying to get in; a gigantic, malevolent wasp, perhaps, stirred from feasting on fallen fruit, or shaken from its nest, clearly madder than hell. It is more than noise, it is palpable vibration, the little brother of a jackhammer's racket—which is the exact truth, because the godawful discord comes from a miniature jackhammer, slowly chipping away at rock.

Such sounds come from the preparation laboratories of vertebrate paleontologists. Only rarely can fossils be plucked from the ground neatly; most have to be chipped, or pried, or blasted out, and when finally removed they usually are still covered by various amounts of obscuring rocky matrix. The task of cleaning the newly discovered bones or shells is then given to one of the world's most specialized professions: fossil preparators. These people spend their lives in noisy, dirty, dusty surroundings, usually in windowless basements, where they slowly and patiently remove fossil prizes from rock. They are unusual people, part artist and part scientist; and almost without exception they love their work, the painstaking exposition of the fossil record. Some arrive into their profession as frustrated, failed scientists, whose love for the fossil record was insufficient in the eyes of some faculty examining board; others, like Jack Horner, take the opposite route, moving from preparator to academic. Some, perhaps, are failed sculptors, for the emergence of a beautiful fossil from a shapeless form of surrounding rock is as sublime a process as any man-made work of art, and oftentimes more beautiful.

My reflections on fossil preparators stem in no small part from my prolonged stay as a guest scientist of the South African Museum in Cape Town. One of the great joys of a university teaching position is a wonderful and, it is hoped, not soon to be extinct institution known as a sabbatical, where seven years' service to undergraduate education earns a year off at partial pay. In late 1991 my first sabbatical from the University of Washington brought me to Cape Town, to a small desk amid giant cabinets filled with fossils and next to a huge room where three women worked eight hours a day at removing the most wondrous Late Paleozoic reptile skeletons from extraordinarily hard rock, rock quarried from the Karroo desert.

The South African Museum is nestled amid the great Dutch East India Company Gardens in the central part of Cape Town; this large botanical garden is more than three hundred years old, and filled with exotic plants of great age and beauty. The museum itself rises harmoniously out of the gardens; stately and Victorian, it harkens back to colonial days. And like other museums from that era, such as the Smithsonian Institution of the United States, or the British Museum of Natural History in England, the South African Museum is a storehouse of great treasures, the designated repository for anthropological, geological, and biological specimens collected over the centuries from South Africa.

Far from the public galleries, past security guards courteous but firm, rest gigantic metal cases neatly arranged. These enormous sarcophagi contain over 15,000 individually numbered skeletons, skulls, and bones from the ancient Karroo, each the result of long prospecting, then patient excavation, finally emerging from the preparator's care to be buried anew, this time in a deep metal coffin. But this new interment is quite different from the old, for an endless stream of scientists arrives from the four corners of the globe to see these ancient treasures, using them to test new ideas and theories about evolution and the history of life; or, as in my case, to test ideas about the history of death. Every year, field parties foray out into the vast desert to seek new treasures, and every year the racket of preparation continues as these newly collected fossils are slowly, patiently removed from their rocky garments, increasing the number of once-living creatures in this giant hall of the dead.

I arrived amid this great boneyard fresh from the killing fields of a younger catastrophe; my stop immediately prior to South Africa had been among the 65-million-year-old Cretaceous-Tertiary boundary sections of Spain and France, areas where the end-Mesozoic extinctions among oceanic creatures can be studied in detail. But after ten years of toil in the Mesozoic graveyards, I was curious to see an extinction very different from that which I have long studied: certainly older, supposedly slower, on land, and among vertebrates. Thus my decision to see the Karroo with its record of protomammals. I would compare the tempo of their extinction to that of the group I have long worked on, the ammonites, whose last fossils tell a story of sudden, catastrophic death some 65 million years ago.

In contrast to the extinction of the ammonites and other victims of the Second Event, where a decade's research by hundreds of scientists around the globe has built up a picture of a mass extinction taking a few thousand to tens of thousands of years at most, all published literature about the Karroo states that the extinction of the protomammals was slow, drawn out, lasting millions of years. But nowhere can the proof of this statement be found, other than in the memories of the Karroo fossil hunters. Nowhere are actual ranges of fossils from measured stratigraphic sections illustrated; in no publication are there photographs or diagrams showing the localities from which these observations had been made. Even more depressing, it is believed that the last several million years of Paleozoic time are not represented by strata found in the Karroo—but the scientific paper asserting this belief provided no source or proof. The scientists simply said, "This is the way it is, trust us."

I had come to South Africa to see the Karroo, not to research it; mostly I had come to Africa to document the ravages of the Third Event, not the First, and to find a peaceful haven to write this book. Much of my literature research on the First Event had been completed prior to my arrival; I had learned that both in the sea and on land, animals (including the protomammals) died out slowly over the last several million years of the Paleozoic Era; the favored explanation for their extinction was that a slow deterioration of the climate, accompanied by a lowering of sea level, created conditions inimical to most life on earth at that time. It may well be that this is exactly what happened. Maybe. But of one thing I am sure: At least for the protomammals, the answers to this riddle of death lie not in any journal yet published or in any extant museum collection. The identity of what may have been the greatest killer in the long history of the earth lies still undetected among the strata and bony corpses slowly eroding out of the Karroo.

4

I sometimes ask undergraduate students to define science. Many reply that science is an accumulation of facts. Others add a time component, suggesting that science is knowledge discovered over time; a few take a bolder stance and equate the word with a human activity, rather than the result of that activity; in this view science is a verb, not a noun. Mr. Webster (depending on the thickness of the edition) talks about all of the above. Science is certainly about facts (and about falsehoods as well). But if you listen to scientists, or watch them in action, they seem to be preoccupied with questions rather than answers. In the absence of a good question, observations or facts capable of being wrenched from Mother Nature are often meaningless. A corollary of this is that the answers usually best fit a given question—and can be very misleading if applied, out of context, to an entirely different question. This holds true for the Karroo. Men and women have been journeying there in search of fossils for more than a century and a half; first in wagons, then in cars, trucks, and helicopters, a treasure of skeletons has been brought out of the Land of Thirst, ultimately ending up in museums scattered across the world. Each of these bones was collected in slightly different fashion, for no two bones are alike, nor are the rocks from which they must be found and extracted. And in an analogous sense, these bones have been extracted from the Karroo's embrace for many different reasons. Some have been acquired as trophies or curios; some to be put on public display. But very few of the Karroo's fossils are of sufficient quality to serve as museum specimens, and most have been sought out for other reasons. Most of the Karroo's vertebrate fossils have been collected by scientists interested in one of two questions: What can this bone tell me about the age of the Karroo? Or, what can this bone tell me about the course of vertebrate evolution? This latter question has generated the most scientific interest, perhaps, and the greatest number of specimens, for the Karroo has long been known to hold the key to one of the greatest scientific questions of all, a query of relevance to all of us: From what sort of creature did our first mammalian ancestors spring?

Deciphering the evolutionary history of the Class Mammalia, mankind's class, has been the holiest of grails, luring generations of paleontologists into the Karroo. But science progresses as questions are answered and new ones emerge; the evolution of our furry forebears is now well understood, and new sets of questions are now more pressing. Sadly, however, data collected to answer one question are often of little use in deciphering another. Such is the case for the giant collections of Karroo bones. They have told us, in superb detail, great stories about evolutionary change. And they have served as reliable timekeepers, producing an accurate chronology of Karroo sedimentation and allowing temporal correlation of this desert with far distant lands. But the collections now housed in the world's museums tell very little about questions now more pressing, and of greater importance: What kills off seemingly stable faunas of land-living vertebrates, such as the one living in the Karroo at the end of the Paleozoic Era? How fast did it happen, and can it happen again? Can the Karroo bones tell us if it is happening in our world, at this moment?

5

Surely the spectral skulls eroding out of the Karroo hills have been long noticed by humanity. What must the various tribes have thought of the white bones and strange, vacant eyes slowly emerging from the gritty strata? The first white man's record of such notice dates back to 1827, when a fossil tooth and then a skeleton of a prehistoric beast were collected from Beaufort West, a town in the heart of the Karroo. This discovery was announced in the *South African Quarterly Journal* in 1831. These discoveries, however, were soon overshadowed by the prodigious efforts of one man, Andrew Geddes Bain, who first demonstrated to the scientific world the great wealth of fossils to be found in Karroo rocks.

Bain was an extraordinary character, and clearly the right man at the right time. A Scotsman who migrated to South Africa in 1816, Bain took up residence in the Karroo and soon became a hunter and explorer of note. When his trading camp was pillaged by Ndebele tribesmen in 1834, Bain barely escaped with his life. He became an officer leading black troops in an ongoing frontier war, and then transferred into the military engineering corps and became involved in road building. In this endeavor he began to stumble upon fossil bones. When he read Charles Lyell's great masterpiece, *Principles of Geology,* he realized that he had found his life's work. Bain soon spent virtually all of his time "fossicking" for fossils. He had to hire a room to keep his burgeoning collection of fossil skeletons, for he could find no institution in the country willing to take them. In frustration, he sent off his collections to England, where they were immediately seized by the two leading anatomists of the day, Richard Owen and Thomas Huxley. I can imagine the two scientists' increasing excitement as they carefully unpacked the South African fossils, finding strange, leering skulls, each bearing a menacing pair of downward-curving tusks. It didn't take Owen, the man who coined the word dinosaur, long to conclude that these ancient Karroo bones represented a group of animals completely unknown until that time. Owen soon described them as the first examples of a new order of reptiles, and later they became key evidence in a great debate about the origin of mammals.

Bain, meanwhile, was busily excavating skeletons at a mad pace. His was the passion of the truly converted. The Karroo, indeed all of South Africa at that time, was a wild and woolly place; this was the time, after all, when the Boers were starting their great treks, and major battles between whites and the native tribes were taking place at scattered localities throughout South Africa. Amid this migration and war, Andrew Geddes Bain unconcernedly went about his business of collecting protomammals. His apparently complete nonchalance to the surrounding chaos foreshadowed that of American paleontologists forty years later, who insisted on searching for dinosaur bones in eastern Montana soon after General George A. Custer was killed by the Sioux Indians, and kept doing so during twenty years of wild west Indian wars. One can only conclude that God in His wisdom certainly looks after the mad race of vertebrate paleontologists.

Bain's rather curious affectation did not go unnoticed by his conservative neighbors and countrymen. The Boers were a paranoid bunch, and had every reason to be. Most had fled Europe because of religious persecution and threats of death, and then were chased out of the rich agricultural land around Cape Town by the English. As they moved into the Karroo, the Transvaal, and Natal, they ran up against very determined opposition from the local tribes, who naturally enough took some umbrage at their grazing land being appropriated by white men citing God's will. In the savage wars that followed, the Boers (and tribesmen) suffered great loss of life. This was not a group of people long on tolerance, or with an appreciation of farce. So when one of their own (Bain had married into a prominent Boer family) began to drop everything to better pursue his beloved, dirty, not even pretty stones, a few fierce eyebrows began to be raised. Bain began his career as fossil collector at age forty. Up to that time it had not been considered at all strange that this white man had served as an officer in an army composed of black soldiers. The Boers could look past the fact that this fierce warrior routinely recited Shakespeare, wrote doggerel, and even had an original play of his performed. But for a grown man to walk around the Karroo digging up bones? Happily, much of this unfavorable opinion was changed radically by the arrival of a check for 220 pounds sterling from Her Majesty's Government—pushed through by Owen and Huxley, in recognition of Bain's great discovery. Now here was something the thrifty Boers could understand—he must be doing it for the money! Fossils became respectable, but, unfortunately, a precedent had been set: Fossils from the Karroo had gained a price tag.

Bain went on to collect many more protomammals, and even founded a family dynasty of fossil finders, for his son followed in his footsteps, collecting skeletons for the newly constructed South African Museum in Cape Town. Meanwhile, the collections sent to the British Museum by the elder Bain languished for many years, their true significance overlooked; Owen and Huxley believed that mammals had originated from amphibians, completely bypassing any evolutionary detour through the reptiles. In this scheme, Bain's fossils were a unique and separate branch of reptiles: interesting in their own right, but a sterile branch of evolution, giving rise to nothing of consequence. With Darwin's Theory arriving as a great bombshell in 1858, however, scientists began to take a more critical look at fossils, and old bones in particular; they were spurred on by a search for the ultimate origin of humanity. In 1870 the American dinosaur hunter and evolutionist Edward Drinker Cope was able to examine a fossil skull from a large dicynodont herbivore named *Lystrosaurus,* one of the most common fossils found in the Karroo. Based on this study, Cope arrived at a conclusion very different from that of Owen and Huxley: He announced that reptiles, not amphibians, were the immediate ancestors of the first mammals, and that the protomammals discovered by Bain came from the first branch of this evolutionary lineage. In one stroke the Karroo fossils went from a sterile sideline to the main event in the search for the root stock of mammals. Following Cope's discovery, a great fossil hunt began in the Karroo, conducted by a variety of interesting characters. But the dominant figure in the hunt for the first mammalian ancestor was a man even more bizarre than his predecessors or contemporaries; onto the broad stage of the Karroo strode a character too bad to be true: Robert Broom.

Broom was a medical doctor trained in Scotland, but his real passion was paleontology. He arrived in Cape Town in 1897 with the express purpose of tracing mammalian ancestry back to the reptiles. Broom had visited Australia, where he had studied marsupials and the odd monotreme mammals—including the egg-laying, duck-billed platypus—in the hopes of deducing the origin of the mammals from the stocks of these extant but primitive forms. But he soon found that only fossils would yield the necessary clues about mammalian evolutionary history.

Broom first supported his fossil hunting by practicing medicine across the Karroo, but after several years he joined a university faculty in a small town near Cape Town. He was a prodigious collector and writer, and soon began publishing a long stream of scientific papers describing his finds, most of them one page long, each describing yet another new species of protomammal. Before long Broom was well known across the Karroo, apparently as much for his bedside manner as his fossil digging. To many of the isolated frontier women in the lonesome farms dotting the Karroo, the appearance of the witty, impeccably dressed doctor must have been a pleasant change of pace, and if the doctor came calling during the heat of the day, when the hardworking menfolk were tending fences and sheep, who could say there was wrong to it? Almost a century later, as I crossed the Karroo, a familiar story seemed to echo down across the decades: Dr. Robert Broom was quite a hit with the ladies. It was a good thing the Karroo was so vast, the fossil-bearing exposures so widespread, for Robert Broom quickly wore out his welcome on more than one farm. As a friend of mine from the South African Museum put it, "Robert Broom jumped out of more than one bedroom window."

Broom must be given his due. His energy in collecting and describing new species laid the groundwork for the Karroo biostratigraphy still in use today, and was a major contribution. Employing the principle of faunal succession first discovered by the English canal builder William Smith, Broom proposed a series of faunal zones that break up the Karroo strata into short time units. Because of the rapidity with which the ancient Karroo protomammals speciated and their warm-blooded propensity for extinction soon thereafter, Broom was able to do this by describing rock units containing key, index species of short duration. This pioneering biostratigraphy was, perhaps, Broom's most lasting contribution. This contribution had its price, however.

By all accounts Robert Broom soon considered himself too large a figure for what he saw as the limited stage of South African science; he clamored for the attention of European and American intellectual life. In the heady days of the early twentieth century, paleontology was one of the most glamorous branches of science; decades of triumphant fossil collection in Europe and America prior to the turn of the century had revealed a rich tapestry of evolution among the vertebrates. The great dinosaur hunters of America and the anatomists of Europe were busy formulating theory and phylogeny at a breathtaking pace, and Broom yearned to take his rightful place among them. He began to see himself as superior to the South Africans, and he let them know it. Nevertheless, in 1905 Broom was named honorary keeper of fossils of the South African Museum and charged with building up the collections. Using his position, he soon began to receive Karroo fossils from many sources, collected and supposedly destined for the South African Museum. But Broom personally kept these collections. In 1909 he traveled to New York, where he briefly studied the fossils of *Dimetrodon* and other finbacks in the collections of the American Museum of Natural History. There he met American paleontologist Henry Fairfield Osborn, curator of vertebrates for the great museum, and there Broom must have first hatched a scheme of treachery that causes him to be reviled still by his countrymen.

Broom's studies confirmed the earlier work of Edward Cope, who had first suggested that Karroo protomammals were younger than the fin-backed reptiles of America, thus making them transitional species between fin-backs and the first true mammals. In recognition of these studies, Broom finally achieved the international acclaim he so craved, and in 1913 he traveled to London, where he gave the Croonian Lecture before the Royal Society, one of England's most prestigious scientific accolades. Henry Osborn traveled across the Atlantic to attend Broom's lecture, and during this London sojourn the two men concluded a business deal. Broom sold a large collection of Karroo fossils to the American Museum of Natural History, receiving 12,000 pounds sterling—a sum large enough to make him financially secure for life. This beautiful collection sailed by steamship across the Atlantic from South Africa and was curated into the vast holdings of the grand old museum on Central Park, where it remains still. The value of this collection is today beyond price. Unfortunately for South Africa, the collection wasn't Broom's to sell: He had sold the protomammal collection of the South African Museum to the Americans.

Such piracy now would put a person in one of South Africa's very unpleasant prisons, and Broom's massive theft may be the main reason that today that country has one of the world's most stringent laws about fossil collection: Unless you have a permit, you cannot touch a fossil in the Karroo. (Alone among industrialized nations, the United States has no protection of its fossil resources. Treasures of enormous value are routinely being taken from the country because of this criminal oversight.) In 1913, however, the world was hurtling into war, and by the time the shooting was over, and people in the know realized what had happened, it was too late. Broom, triumphant and unrepentant, was fired from his position in the South African Museum and barred from entering there, but otherwise emerged unscathed and far richer. To this day South African science mourns the loss of its great collection.

Robert Broom continued to collect and describe fossils from the Karroo following his great treachery. But he had lost that most valuable commodity, trust. The keepers of South Africa's other museums began to take care when he arrived to study collections. He was usually assigned an assistant, if only to keep watch over him. According to a current staff member of the South African Museum, Broom was not adverse to pocketing a specimen now and then, or performing his own crude preparation on the spot to better expose some feature of a bone that interested him. His list of publications continued to grow as each new skull became a new species, but this practice began to fill the literature with many invalid new names, for he defined new specimens belonging to already described species, making the Karroo fauna look far more diverse than we now know it to have been. Paleontologists are still working to rectify Broom's bizarre taxonomic practices.

Broom labored mightily for fifty years, with unquenchable certainty, never doubting his superiority. But energetic as he was, he could not forever keep other scientists out of the Karroo. During the first half of the twentieth century other collections were gathered, by museums in Johannesburg and at the Geological Survey in Pretoria, and again at the South African Museum in Cape Town. But perhaps the most important yet most curious of collections was gathered by three generations of one family living in the heart of the Karroo.

Sidney Rubidge, the founder of this paleontological dynasty, was born in the small Karroo town of Graaf Reinett in 1887, and at a young age took over a large sheep ranch known as Wellwood Farm. The name of this ranch seems a cruel joke, for at its founding in 1840 it had neither a well nor wood. Located in a dry valley surrounded by high walls of fossiliferous Karroo strata, the Wellwood ranch site was less desirous than most in the area, for all water had to be caught from the fickle rains in large cisterns or brought in with wagons. Nevertheless, by the time young Sidney took it over, the ranch was a thriving concern, known throughout the Karroo for the excellence of its merino sheep.

Like Andrew Geddes Bain, in midlife Sidney Rubidge was bitten by a deep and overwhelming passion: fossils. He began to collect fossil bones from the hills around his homesite, which are particularly rich in protomammal skeletons of very large size. He imparted his great love of fossil hunting first to his son and then down to the next generation as well, among his grandsons. But unlike many private collectors who eventually realize the value of their collections to science and turn them over to museums, the Rubidges held on to their material and even began hiring local people to collect more for them. Robert Broom soon learned of this great fossil bonanza accumulating in the heart of the Karroo, and cultivated old man Rubidge.

Broom began to make regular visits to the Rubidge farm. True to his nature, he wrote scientific papers about the newly collected specimens he found awaiting him on each successive trip, almost invariably describing new species. But instead of following the normal, scientific practice of placing the best fossil examples of his various new species into one of the established museums, so that all interested scientists could study them, Broom left the crucial fossils at Wellwood Farm. The specimens were excavated, cleaned, and then put on display within the growing, private Rubidge Museum. By the time of Sidney Rubidge's death in the middle part of this century, the museum housed 840 skulls, of which an astounding 117 were type specimens of new species defined by Broom. The type of a species is the most cherished commodity of any museum, for it is the specimen, or specimens, best showing the characteristics defining the species in question. Type specimens are usually the best preserved or most complete specimens of a species ever found. Any scientist studying systematics and taxonomy (the science and practice of classification of organisms) must spend a great deal of time in museums studying type specimens.

Robert Broom was a notorious "splitter"—his concept of a species was so narrowly defined that any slight difference in morphology was sufficient cause, in his mind, to justify his definition of another new species. Paleontologists studying Karroo protomammals subsequent to Broom have shown that the majority of his species are not valid—they are simply variants of previously defined species. Nevertheless, the sheer number of new names Broom introduced ensured that some are valid, with their type specimens today sitting in glass cases in the middle of the Karroo. This fact is a source of no little embarrassment for South African paleontologists, since any scientist interested in understanding the nature of the Karroo protomammals must journey to Wellwood Farm. It seems the ultimate irony that such a pilgrimage is necessary, in a country that now allows no private fossil collecting, for there can be no more isolated museum in the entire world.

Miles from anywhere except small Karroo towns, its collections neatly arranged in a small house amid a sheep farm, the Rubidge Museum today attracts a few curious visitors and an occasional, determined scientist. I made this trip out of curiosity rather than need, and was wonderfully welcomed by the current generation of Rubidges running the place. Following tea and much discussion about the current drought and price of sheep's wool, I was ushered into the neat, white house containing the fossils, where I beheld its inhabitants with amazement. The number and quality of protomammal fossils held in this white-

Dr. Roger Smith, my guide to Karroo geology, seen amid fossils found in the Rubidge Museum.

washed room is beyond compare. All four walls are lined with large glass cases, each filled with the skulls and skeletons of the Karroo's ancient inhabitants. The huge, silent skulls are the most affecting; dark dusty orbits peer blindly into the room from most of them, but in other skulls some Rubidge joker has put painted glass eyes into the empty eye sockets, giving a most lifelike expression to the grinning skulls. It was very creepy. Pausing from my inspection of these gorgon heads, I looked elsewhere in the room, to see numerous framed pictures. Many held photographs of various Rubidges, but the most common image staring out from these pictures was the supercilious, smiling face of Robert Broom, smirking, it seemed to me, at having made but another in a long line of inferior scientists come to him, to pay homage to a priceless collection buried in obscurity. He made me very angry that day.

Broom and his generation of paleontologists searched the Karroo for one reason: They longed to find the true ancestors of the mammals. Like Burton and Speke on their search for the source of the Nile, they followed fossil lineages back through the rivers of time, into ever-narrower evolutionary channels, longing to find the ancient, first trickle of creatures that have so transformed the earth. Broom died before reaching the source of the mammals. The first fossil universally acknowledged to be from a true mammal was recovered in 1960 from Triassic strata in the Karroo, nine years after Broom's death.

Like Broom and Bain, most of those excavating fossils out of the

Karroo have not been trained geologists. Most were anatomists of one sort or another, forced into geological field work by necessity. Many, such as the legendary collector James Kitching of Witwaterstrand University, have become excellent geologists in their own right. But none of these collectors worried overly about the locations at which his fossils were found. Looking over the locality records for the numerous fossils stored in the various South African museums, I found that most were located only by a given farm or valley. In no case could the original sites of fossils collected prior to the 1980s be relocated or plotted on a map or aerial photograph. Because of this imprecision, there is no way that we can use current protomammal collections to arrive at any information about how fast they went extinct during the First Event. Nevertheless, without exception, every learned source writing about the protomammal extinction in the Karroo states that it was a long-drawn-out affair. What a load of nonsense. No one has yet completed the work necessary to make any sort of statement about how fast the extinction was. But from what I have seen, I would be willing to wager a great deal that the Karroo protomammals died out with stunning rapidity, in a million years or less. To an observer fresh from the graveyards of the quick, catastrophic Second Event, the First looks hauntingly familiar.

6

Here is the favored scenario, found in all texts and references dealing with earth history: By the end of the Paleozoic Era, some 245 million years ago, the Gondwana continental assemblage had moved northward to crash into a northern supercontinent, composed of a united North America, Europe, and Asia. The resultant collision created the largest single continent in the history of the earth. This continental amalgamation coincided with the onset of the First Event. Two effects of this gigantic, tectonic embrace supposedly produced the extinctions.

First: The earth's climate changed. The interiors of this gigantic continent grew hotter in summer and colder in winter. Because of its immense size, huge areas of this supercontinent could no longer be cooled or warmed by steadying, maritime influences. Temperatures in summer would have climbed well above 100° F each day, then plummeted each night. During winter the opposite would have occurred: Freezing, dry

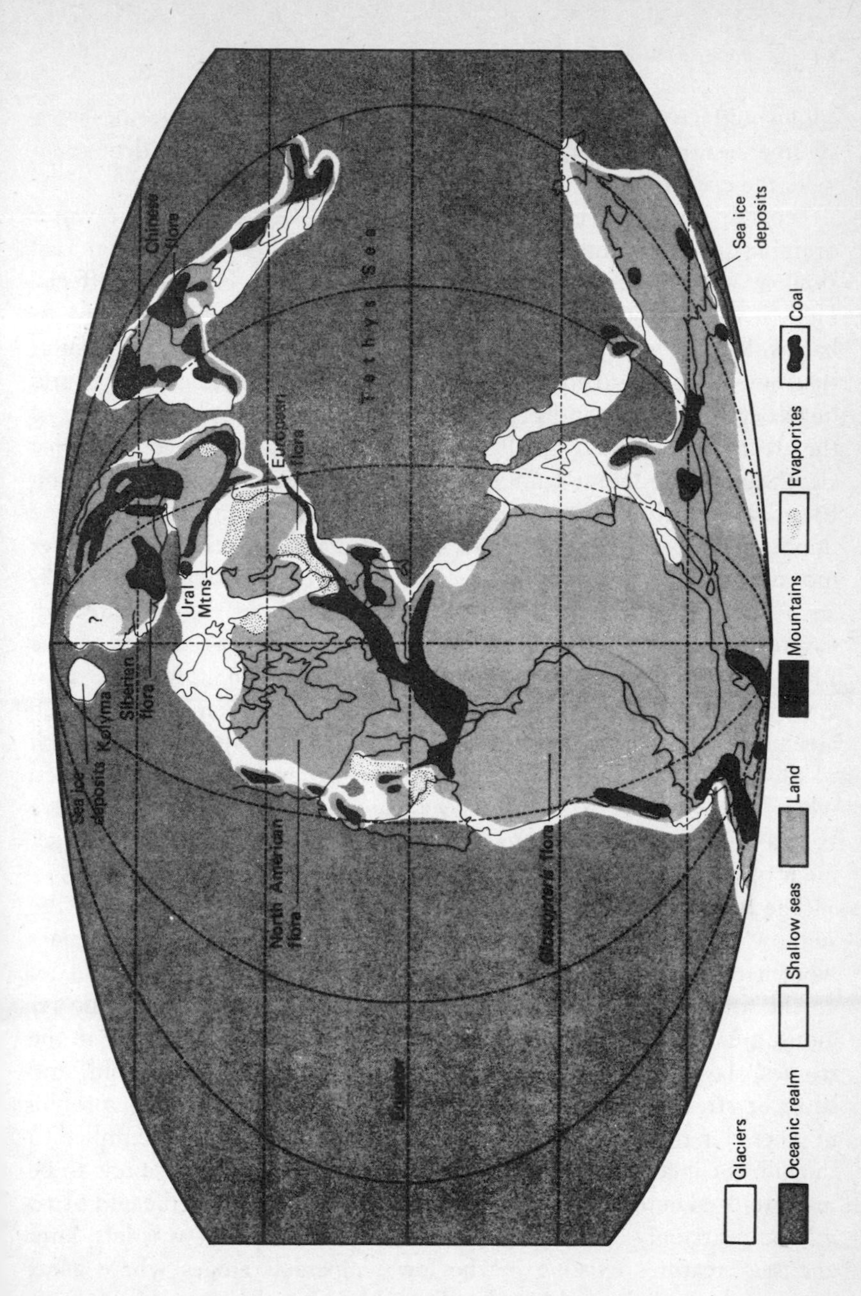

Continental configurations at the end of the Paleozoic Era, showing all of the continents merged into one giant supercontinent.

cold would have prevailed over much of the interior. The extremes—the summer heat and winter cold, perhaps accompanied by great drought—were the executioners.

Second: All over the earth, the level of the oceans fell. Like water draining out of a bathtub, the oceans dropped as much as 300 feet. This slow yet inexorable drop in global sea level had two immediate effects. First, it caused the coastlines to expand as the sea retreated into its basins. Second, wide, interior seas found on virtually every continent at that time drained and emptied, leaving only poisonous, briny lakes and huge deposits of salt and gypsum in their place. The disappearance of the great interior seas of the Paleozoic is thought to have been the most significant single factor leading to the marine extinctions of the First Event. The earth at that time was quite different from today's in one important respect: The continents were far flatter, for there were fewer mountains. The major mountain systems on the earth today—the Rockies, Andes, Himalayas, Cascades, and Alps—had not yet been created. Continents were often covered by wide, shallow seas similar to the Black Sea of today, but far larger. Within these shallow seas, most Paleozoic marine life lived. Many of these seas had been havens to rich assemblages of marine life, but as they receded from their basins, much of this life was extinguished. The drop in sea level may have occurred when large volumes of ice grew over the polar regions, or it may have been a side effect of continental drift, caused by a slight volume enlargement of the ocean basins as the great continents coalesced.

The two processes were linked. As the climate grew more arid, the shallow seas evaporated all the faster, and with their loss the climate worsened, for these great inland bodies of water must have had an ameliorating influence on the climate. In the earth's high-latitude regions, great glaciations stalked the land, while huge deserts grew in the tropics, slaying forest and fauna alike. The decades of heat, cold, and drought stretched into millennia, and then the millennia themselves numbered in the thousands. The earth's species gradually succumbed to the killing climate, slowly falling away like browning leaves, a few to be immortalized in rock, the rest to pass from all memory. By the end of 10 million years only a tiny percentage of Paleozoic species was left, land and sea creatures existing in the few temperate refuges where great equatorial heat balanced frigid polar cold. Or so the story goes.

Something seems to be missing between this neatly prescribed cause and effect. There is certainly no doubt that climate change can cause extinctions. But can climate change alone be the executioner of more than 90 percent of all species on the earth? Although the great continental amalgamation occurring at the end of the Paleozoic Era was unique, during other periods in the earth's long history, climate has changed and sea level has dropped without attendant extinctions.

The long-held assumption that the First Event was of great duration began to unravel in 1988, with an obscure publication by Polish geologists, who described a most peculiar geochemical change in fossils found in Paleozoic rocks from the Alps. These scientists sampled the shells of brachiopods, shellfish common during the Paleozoic but largely extinguished by the First Event. By analyzing the shell chemistry of these fossils, collected from a thin stratal succession, the scientists discovered a very peculiar chemical change. The chemicals examined were carbon and oxygen isotopes derived from the calcitic shells of the fossils. All elements have several isotopes, and the ratios of the various isotopes have proven very useful in studying ancient rocks. Many such measurements are used as rock chronometers, the most well known being carbon 14, which, when compared to the amount of "normal" carbon, C12, yields very accurate age estimates for organic material produced during the last 50,000 years. Carbon has another isotope, C13, which allows estimates of a very different sort. The ratio of carbon 13 to carbon 12 tells much about organic productivity on land and in the sea. When productivity is high, meaning that a great deal of carbon derived from the atmosphere or ocean is being taken up by plants and turned into living tissue, the ratio of C13 to 12 is high; when little photosynthesis is taking place, the ratio lowers. Since these carbon atoms are taken up by living tissue, and sometimes preserved in bone and shell, they can be sampled long after the living creature containing them has died.

The Polish scientists examined this ratio, and their findings were curious. Perhaps a million years before the end of the Paleozoic Era, a sudden spurt in productivity occurred, followed by a decline so profound that it speaks of oceanographic change and a virtually unprecedented loss of oceanic productivity. If the figures are accurate, it suggests that most of the plankton in the seas died out. As the plankton died, so too did all of the creatures up the food chain. The problem with this study is that it was necessarily incomplete. Because no marine strata straddle the Permo-Triassic boundary—an interval is missing in every location yet examined—the most critical part of this story has not been sampled. Another problem is that the investigators could not arrive at an accurate estimate for the duration of their observed event. In their view, it could have taken place over a period as long as a million years, or less—perhaps much less. In a recent summary of this work, paleontologist Anthony Hoffman described this occurrence as "The most profound oceanographic event so far observed during the last 560 million years." In fact, such an isotopic shift has been seen on this scale at only one other time in earth history: at the Cretaceous-Tertiary boundary transition, the time of the Second Event.

In 1990 geologists from South Africa conducted an analogous study, looking at carbon isotope values derived from the teeth of Karroo protomammals. To everyone's great surprise they found an astonishing parallel to the results from the earlier marine analyses. Carbon isotopes showed a brief rise, followed by a great, sudden fall. But the implications of this eerie similarity were not much appreciated, because of the universally held view that the Karroo, like all known marine Permo-Triassic sections, has a long, missing interval at the critical time when the extinctions must have been at their height. The isotopic shifts found in the Karroo were thus thought to have occurred well before the event in the sea, and were therefore simply coincidental. But what if this long-held view about the Karroo is wrong? What if there is no missing interval there? This is what I had seen on my second Karroo visit. In the company of Dr. Roger Smith, a specialist on the ancient river deposits comprising the majority of Karroo strata, I had searched for any sign suggesting that a several-million-year break in sedimentation had occurred. None could be found. The implication of this is straightforward: The isotopic shifts—and the extinctions—seen in the sea could have taken place at the same time as did those on land. It would mean that a great disturbance visited the earth some 245 million years ago, simultaneously—and catastrophically—altering the marine and terrestrial biotic systems. It may have been an asteroid strike, as yet undetected because of the dearth of stratigraphic sections recording the critical time interval, or it may have been some other, still-unnamed demon reaping a grim harvest. Whatever its identity, the cause of the First Event will not remain anonymous much longer. We know where to look; apparently only the Karroo holds the critical moment of time when the killer walked the land. And we know how to look: Foot by foot, collecting everything across the critical stratal interval, sampling for plant pollen and trace fossils, sediment grain size, and protomammal skeletons; by collecting everything in great detail in the Karroo sections we will find out how long the First Event took, and what it actually killed; we will find if the earth's creatures all died out together or sequentially. And when we know this, we will have solved a great, long-running murder mystery. I hope to be there when the killer is finally unmasked.

7

Head jammed against the aircraft's window, I surfaced painfully from uneasy dreams. The faint, luminous dial on my watch said 5 A.M., meaning that I had been flying for nine hours since leaving Cape Town, and should be only three hours from London. The cabin was dark and, all around, my fellow passengers were in various states of uncomfortable sleep. I wondered what had awakened me, and then felt my ears pop; the plane was descending. This was quickly confirmed as the captain of the South African Airways 747 ordered seats up and seat belts on. Through the window, I saw a few lights punctuating the blackness below. Trees rushed up at me, and we pancaked suddenly onto the runway. Finally rolling to a stop in front of dilapidated Quonset huts, we were welcomed to Abidjan, capital of the Ivory Coast, for a very unscheduled stop.

And there we sat. South Africa, pariah among African nations, is not allowed to fly its national airline over most of Africa, let alone land anywhere. We had made an emergency landing, grudgingly permitted by this poor, equatorial country. Furious negotiations undoubtedly were taking place in the cockpit as the captain bargained for fuel and minor repairs. Annoyed, bored, dog-tired, and still facing hours of flight time before reaching London, I rummaged around in my bag for distraction and amusement, eventually finding a tissue-wrapped object. I removed the shroud to uncover the tiny stone skull of a Karroo protomammal, a gift from my South African colleagues. Inscrutable and long dead, the tiny head stared blankly at me. "What killed you?" I wondered for the thousandth time, and again had no answer. Was it long, slow climate change, or were you, like your dinosaurian successors, killed by some asteroid falling from space? Either of these possibilities is fine with me. I can deal with slow climate change, and I have no problem with rocks from space, for neither are forces that mankind can have any control over. But what if the protomammals and all the rest of the Paleozoic world were killed off by something else, something applicable to our world, and possibly currently under way? My greatest fear, for our world, is that an increased rate of extinction can eventually reach some threshold point, triggering a cascade of mass extinction, a free-fall of death. Each species on the earth is like a tiny piece in a four-dimensional jigsaw, interlocking with other species and a tiny conducting part of the energy flowing through the living world. But what if species are also stacked together like a giant house of cards, each supporting other species in some small (or large) way, so that if enough species are kicked out of place by their extinction, the entire house falls down? Did that happen at the end of the Paleozoic? Did enough species get killed off to bring down a sudden torrent of extinction, eventually removing 90 percent of the earth's creatures? And how far from that cliff are we today?

My cynical reverie was broken by much loud shouting from the front of the plane, followed by the captain's smooth voice advising us that the cabin will be sprayed with insecticide by officials of the Ivory Coast, to make sure that no undesirable South African organisms invade their proud, equatorial country. With evident glee two large men moved through the cabin, spraying the passengers. As the official on my side of the plane passed by, I pointed my small skull at him, its small, 245-million-year-old death's head glaring fiercely. I received an extra dose of DDT for my trouble. All around me small creatures are undoubtedly dying, as the cabin fills with the noxious spray. There is no escape, no amount of breath-holding sufficient to outwait the chemical barrage. Soon there is only one species of animal left alive in the plane.

Part Two

The Second Event

Chapter Four

Dawn of the Mesozoic

I

The world emerging from the First Event may have been like a landscape in the aftermath of a forest fire. At the end of the conflagration, the burned land is barren. But the seeds of renewal lie everywhere, either in the burned landscape itself, or providentially brought into the charred blackness from refuges that had escaped the brunt of the flames. Soon these seeds burst into life. At first only fast-growing weeds mark the renewal, but gradually a complex series of communities begin to succeed one another, successions of species growing riotously in the absence of old competition and old rules of order. Eventually a climax community of tall trees emerges from the weeds below, shutting off life-giving light to the forest floor and, in the process, dooming the short-lived weeds. Several centuries after the fire, the forest is restored, usually looking little different from its prefire state.

In the immediate aftermath of the First Event, the land and sea were also nearly empty, deserts of little diversity. Soon, however, the forces of evolution and immigration began the process of recovery. New species evolved at a rapid rate, for in an empty world there is very little competition, and virtually any body design will flourish—for a while. Other creatures emerged from refuges, places like the deep sea, perhaps, which seem best in avoiding the ravages of mass extinction. But the analogy between reforestation following a forest fire and the repopulation of the earth following a mass extinction breaks down here, for unlike a forest several hundred years after the fire, the restored earth and its ark load of organisms never comes back with the same, preextinction composition. The clearing of species by extinction opens the faucets of evolutionary change, allowing prodigious bursts of evolutionary experimentation and diversification. Just as baseball is an activity rather than simply an aggregate of willing players, ecosystems are made up not only of creatures but by the way they interact, the nature of energy flow through the system. With a new suite of creatures, energy can flow through the ecosystem in ways far different from those in a preextinction system. Mass extinction not only changes the players, it changes the very rules of the game.

2

At the center of every town and village in France and Britain sits a monument to those fallen in the Great War of 1914–1918. Carved in a variety of stone and each decorated differently, all nevertheless have a unifying component: a list of names of the dead. Never before in the history of mankind had such a slaughter occurred, and as the weary survivors returned to their villages, home from the horrors of the trenches, they discovered that their actions, and the incredible slaughter that resulted, had irrevocably changed the world.

Stony, stratal monuments to the First Event, the great mass extinction closing out the Paleozoic Era, also hold the lists of the dead, engraved not as names but as the fossilized bodies of the long-dead victims. In every outcrop of 245-million-year-old sedimentary rock the same message of mass death can be seen. But not only the roll of fossils demarks this watershed event in the history of life, for the rocks themselves seem to cry out to us. The earliest rocks of the Mesozoic Age are called the Triassic System. They are easily recognized virtually everywhere on the earth because of their distinctive color: bright red. It is as if the earth, having killed off most animals in the First Event, itself became stained by the blood of the dead.

3

The rickety stairway looked westward over a wide valley filled with old wooden houses, today a crack-house ghetto but in 1956 a Seattle chapter of the American Dream. Below me, in the distant streets, marvelous cars cruised by in stately splendor: Buicks, Fords, Chevys, Studebakers, and an occasional finned Caddy going nowhere in particular in the darkening twilight; all American, only American, V-8 two-tones and soft tops murmuring sweetly on 15-cents-a-gallon Richfield ethyl. I loved to watch the great old cars pass by on warm summer evenings, but on this night it was the sky that kept me from bed. My brother and I, already in pajamas, watched this darkening sky and the bright red beacon staring downward from it. In that year the planet Mars made a historically close approach to the earth. Unblinking, a stark crimson lamp, it put to shame the weak and shimmering stars around it. I knew nothing of H. G. Wells, or Gods of War, but I knew that the bright planet in the sky was something extraordinary, for nowhere in my young life had I ever seen such a color. First through binoculars and then through increasingly larger telescopes I revisited Mars in later years, wondering often at the source of its rich red color, a hue burned into my memory. Much later in my life, American technology threw machines onto the Red Planet and sent back pictures of a barren desert world, a wasteland of red sand and rock. But by then I had already walked red deserts, among rocks called the Chinle and Moenkopi formations of Colorado, Utah, and Arizona. My first view of the rich red strata in southwestern Colorado soon after my twentieth birthday was a moment of blissful surprise; like unexpectedly smelling a perfume evoking a long-past, forgotten romance, my first sight of Triassic strata took me immediately back in time to my childhood sighting of Mars, so powerful and similar are the shades of red. Sooner or later our species will make the trip there and finally walk on a planet other than our own, but for the geologists of our planet such a visit will be anticlimax. Even in civilized Connecticut, as well as the wildness of the Karroo and Kazakhstan, the red rocks created in the earliest millennia of the Mesozoic Era seem eloquent testimony to the aftermath of mass extinction, for on Mars and the earth the causes of these red sediments are virtually the same: They speak of erosion and wind, heat and killing cold, the rusting oxidation of sediments deposited in the absence—or, in the case of the earth after the First Event—the scarcity of life.

4

A world ended 245 million years ago. But its replacement took a while to build. Following the First Event, the oceans and land once again filled with creatures, but the filling took many millions of years. In the sea, the great Paleozoic fauna of shelled brachiopods and stalked crinoids, archaic fish and tetracoral reefs, gaudy ammonites and segmented, creeping trilobites, all had been nearly or completely swept away. During the first 10 million years of the Triassic Period, these extinct creatures were gradually replaced by a benthos dominated by clams, snails, and coral reefs not dissimilar to those of today. But the changes in the sea were minor compared to those on land, where the great, Late Paleozoic empire of the protomammals had been almost entirely swept away.

Fossils of fifty-five land vertebrate genera, most belonging to the protomammals, but with some amphibians and early reptiles as well (including the ancestors of turtles, lizards, crocodiles, and dinosaurs) are known from strata deposited immediately prior to the First Event in the Karroo. Only five of these creatures are found in strata following the mass extinction, rocks assigned to the Triassic Period. The story is the same anywhere on the earth where earliest-formed strata of the Mesozoic Era are still preserved. A great dying had emptied the land ecosystems of their animals, and only a scarce few survived.

One would imagine that any group surviving the First Event must have had very special genes indeed. When so many other species perished, what was it about these few survivors that allowed them to pass this great filter in the history of life? Dr. David Jablonski of the University of Chicago has pondered this question, and suggests that those species with wide geographic distributions have a better chance of surviving a mass extinction. Darwin himself suggested that long-lived species, or "living fossils," as he called them, were forms that could tolerate a wide variety of environmental conditions, or lived in habitats where little competition with other species existed. But there is a further possibility—that some species survived the First Event, or any of the other mass extinctions, purely through luck. If this latter assumption is true (and most paleontologists subscribe to this view, so eloquently described by Stephen Jay Gould in *Wonderful Life*), it means that if the history of life with its many millions of species arising, living, and then going extinct were somehow rewound and then replayed, it would never repeat itself in the same fashion. Of the fifty or so land vertebrates found in the youngest, pre–First Event strata, which five would get through a second time? The second time around, would our lineage survive?

Unlike land animals, so grievously killed by the first great extinction, plants appear to have been less affected by the First Event. Although we have no idea how many plant species actually died out during the First Event and how many lived (identification and naming of plant species from this long-ago time is still under way), it appears the event itself did not change the makeup of the plant communities a great deal. Much of the land area on the earth appears to have been dry, cold desert, not dissimilar, perhaps, to the Gobi Desert of Mongolia today. Where humid or even swampy conditions existed, however, forests of ferns and conifers appear to have dominated the landscape. From such settings, we find the most abundant fossil vertebrate of the earliest part of the Mesozoic, a curious creature named *Lystrosaurus*.

Lystrosaurus must be considered a strong candidate for the earth's All-time Ugliest Animal Award. About the size and approximate appearance of a large pig (a comparison probably unfair to pigs), the lystrosaurs seem to have existed in huge numbers at the start of the Mesozoic. They preferred wetter habitats and must have been ponderous, slow-moving (and probably slow-witted) plant eaters not unlike water buffalo. Their bones are found in vast numbers in the Karroo, India, Russia, and Antarctica, a fact causing paleontologists to speculate that these odd herbivores lived in giant herds. They were certainly the most populous land vertebrates in the earliest millennia following the First Event.

Although lacking looks and surely charm, the lystrosaurs appear to have been among the luckiest creatures ever to have lived, for two reasons. First, they (or their immediate ancestors) survived the First Event, beating one-in-ten odds in doing so. (Imagine a revolver with ten bullet chambers, nine of which are loaded. Spin the cylinder, put the barrel in your mouth, and pull the trigger. *Lystrosaurus* got the empty chamber.) But the lystrosaurs had far more than mere survival to be thankful about; following the First Event they found themselves in a world devoid of large predators. Perhaps the oddest aspect of this earliest Mesozoic world, as deduced from the fossils found in the gritty, red, Triassic strata, is that it was without carnivorous land creatures larger than a squirrel. The lystrosaurs roamed fearlessly on the bleak landscape, masters of the earth. Like we humans, they were about the only creatures in the planet's long history to die commonly of old age, rather than in agony from the predator's tooth and claw.

Like the lystrosaurs, and for largely the same reason, the vast majority of 5.5 billion humans now on the earth will go through their entire lives without once worrying about being eaten by a carnivore. To me it is the surest sign that we have entered the bleak winter of mass extinction, for virtually every creature that has ever lived during the last billion years of life on the earth has had natural predators. Our species certainly had them in the past. But the great cats and wolves, which surely have eaten more *Homo sapiens* than any other extinct or living species, are gone or going, and those few humans now being eaten by the occasional crocodile or great white shark are generally succumbing to very bad luck. (Apparently some science fiction writers and Earth Firsters want to genetically invent or reintroduce some man-eating predator to help stabilize human population size. Some variant on the vampire theme seems favored. But it hardly seems a successful platform from which to win public office.) We humans have arrived at this happy(?), predator-free state through our traits of great savagery and fecundity. The lystrosaurs, on the other hand, simply lucked into a world where all of the large predators had been killed off by mass extinction.

By my calculation, there are approximately 250 million tons of living human flesh currently on the earth. Given "normal" evolutionary rules, such a tempting, unexploited resource should produce a whole suite of newly evolved predators. But would our species sit back for the thousands to hundreds of thousands of years that would probably be necessary to evolve efficient, human-eating predators? Would we passively watch, generation after generation, as our future predators' size, intelligence, and ferocity increased until we too were once again among the hunted, rather than our planet's supreme hunters? Ludicrous, of course. But this has been the normal course of events in the world. Predators usually evolve to meet the challenges of eating specific prey. The Early

Mesozoic lystrosaurs didn't have the brains to realize that the small lizard- and house cat–size reptiles nipping at their ankles would, within a million years following the end of the First Event, evolve into large, efficient, *Lystrosaurus*-eating carnivores.

Nature has always seemed to abhor empty ecosystems; the Early Mesozoic world again filled with creatures through the process of new species formation. On land, four survivors of the First Event played major roles in this great adaptive radiation. Lystrosaurs represented one stock, arising from the dicynodont lineage of protomammals that had been the dominant vertebrates of the Karroo and elsewhere on the earth during the Late Paleozoic, prior to the First Event. The amphibians also survived, and flourished in the early part of the Triassic. The third group gave rise to turtles. The fourth group was also composed of reptiles, but of a lineage very different from the turtles and protomammals. This group, known as the diapsids, became the root stock of crocodiles, birds, and dinosaurs.

At a place called Lootsburg Pass, in the middle of South Africa's Karroo desert, the bones of these ancient survivors can be found in the brick-red strata. The remains of the lystrosaurs are the most common by far, and the largest fossils as well. But other treasures occasionally can be found. The remains of giant, probably ferocious amphibians are present, as are even rarer treasures, such as the fossils of two tiny but otherwise quite different reptiles. The sharp teeth found in their skulls suggest that both of these diminutive creatures were carnivores, probably preying on insects and the babies of other land vertebrates. One of these small reptiles was, like *Lystrosaurus,* from the protomammal lineage, and our distant ancestor, for it ultimately gave rise to the mammals. The other was the ancestor of the dinosaurs. To the ponderous lystrosaurs living their blissful, predator-free existence on the muddy riverbanks of this region, some 240 million years ago, these tiny reptiles were surely of no import. This situation ultimately changed, however.

At that time two great lineages of reptiles wrestled for domination of the land: the protomammals and the archosaurs. The protomammals gave it a good try. During much of the Triassic they remained the dominant herbivores of the land, and became relatively successful carnivores as well. *Lystrosaurus* soon became extinct from a combination of competition from more advanced protomammals and from predation by newly evolved carnivores, such as the doglike cynodonts. (This latter group must have looked decidedly canine and may have hunted in packs, much in the manner of modern-day wolves.) But unlike the long stretches of time prior to the First Event, when the Karroo and other land areas were dominated only by protomammals, the Early Mesozoic world began to be filled by increasing numbers of archosaurs, the ancestors of the dinosaurs—and, finally, by the dinosaurs themselves.

5

By Late Triassic time, some 220 million years ago, the giant supercontinent of Pangea began to disintegrate. This largest of all continents had formed in the Late Paleozoic Era when northward-moving Gondwanaland had crashed into the combined North American–Eurasian assemblage. For almost 50 million years this great supercontinent traveled majestically over the globe. Eventually, however, the great subterranean forces that had melded all of the separate continents into one block began to tear apart their handiwork. The separation of this huge continental assemblage commenced with the splitting apart of North America from a combined Europe and northern Africa. This rifting formed the Atlantic Ocean as we know it today.

When continents collide one result is compression, producing a linear mountain chain. When continents split quite different geological features form. Continental disaggregation is brought about by tension or a pulling-apart of the two sides; as the two huge blocks of crust start to pull apart, big linear cracks appear, initially spewing forth enormous volumes of magma welling upward from deep within the earth in the process. Gradually the volcanic activity ceases, and the cracks that were created become wide valleys. Vast quantities of sediment begin to fill the rifts, for the floors in these "rift valleys" are often lower than sea level. Rivers and marshy areas quickly form within the rifts, often becoming rich oases of life. The Great Rift Valley in eastern Africa is a rather recently formed example of this type of feature; the great north–south running valleys of eastern North America, such as the Hudson and Connecticut River valleys, are others. The latter two were formed in the Late Triassic as North America and Europe split apart and then sailed away from one another. More than 200 million years ago, in these wide, tropical valleys, untold numbers of four-legged and two-legged reptiles gathered in the swampy marshland and flowing streams to drink, feed, and breed. Their feet sank into the red muddy shores and streambeds in the process, leaving behind a rich record of footprints and trackways.

The Connecticut River Valley is enormously long; it not only bisects the state it is named after, but the entire state of Massachusetts and part of New Hampshire as well. The Connecticut River runs through a wide valley for most of its course, and when it finally empties into Long Island Sound, it disgorges untold tons of red sediment eroded from the underlying rock during the river's long journey. The red and brown mud, silt, and sand is all eroded, Triassic rock.

If you journey through this valley, you can note several interesting facts. The rocks making up the broad valley show the characteristic red color of the Triassic, but they are not exclusively sedimentary rock. Interlayered with the softer, eroding sedimentary strata are thick deposits of hard lava as well, stretching, like the valley itself, in a nearly perfect north–south line. The great palisades of ancient lava often hold up the sides of the wide valleys in the region, with the softer sedimentary rock found in the valley floors. Even more peculiar, if you wander close to one of the rocky sandstone ledges or shaley stratal sheets exposed by the meandering river, you might find numerous footprints in the brown or red rock. Looking much like bird tracks, this graffiti comes down through the long ages from the ancient Triassic world.

The footprints in the Connecticut valley's red rocks have long been noted; by the end of the eighteenth century they were commonly known as Noah's raven tracks, and the three-toed, avian shape of most suggested to someone that the footprints had been made long ago, by ancient races of gigantic birds. This was the explanation of Edward Hitchcock, professor of natural history and president of Amherst College during the mid-1800s, the scientist who first studied and collected the Connecticut River Valley tracks. Hitchcock first became interested in the innumerable footprints to be found in the valley in 1835, and devoted the rest of his life to their study. Eventually he had a museum built to house his huge collection of footprints. Hitchcock was clearly a paleontological time traveler, although his time machine gave him the forgivably false vision that he was dealing with giant bipedal birds, judging from the following passage:

> Whatever doubts we may entertain as to the exact place on the zoological scale which these animals occupied, one feels sure that many of them were peculiar and gigantic: and I have experienced all the excitement and romance, as I have gone back into those immensely remote ages, and watched those shores along which these enormous and heteroclitic beings walked. Now I have seen, in scientific vision, an apterous bird, some twelve or fifteen feet high,—nay, large flocks of them—, walking over the muddy surface, followed by many others of analogous character, but of smaller size. Next comes a biped animal, a bird, perhaps, with a foot and heel nearly two feet long. Then a host of lesser bipeds, formed on the same general type; and among them several quadrupeds with disproportioned feet, yet many of them stilted high, while others are crawling along the surface, with sprawling limbs. Strange, indeed, is this menagerie of remote sandstone days.

I was delighted by this passage, written at a time before any reconstructed skeletons of dinosaurs were known; I admit, however, to having been stumped by the words heteroclitic and apterous. Happily, Mr. Webster straightened out the passage: Hitchcock was describing unusual, wingless birds of great height. This is as good a description of a dinosaur as any.

Unfortunately for the professor, no bones emerged during his lifetime to solve the identity of his heteroclitic, apterous birds. Although rich with tracks, the red sandstones and shales were almost devoid of bones, so the origin of the trackways remained a mystery. The good professor went to his grave still thinking that those three-toed footprints came from large birds. We now know that the tracks were made by the ancestors of birds—the first dinosaurs—as well as by other early reptiles such as thecodonts, phytosaurs, and protomammals, which all lived on the muddy riverbanks of the Connecticut River Valley region more than 200 million years ago.

During the nineteenth century and the early part of the twentieth, the maroon sandstones of the Connecticut River Valley were quarried for building stone. The resulting product produced an architectural icon, for the "brownstones" of New York City and elsewhere along the eastern seaboard of North America are entirely constructed of Triassic sandstones, and more than one of these buildings surely holds the bones or footprints of the ancient inhabitants of the Connecticut River Valley.

6

While the eastern coast of North America was being born some 200 million years ago through the rifting apart of two continents, the western coast began to feel the first pangs of mountain building that would eventually result in the Rocky Mountains. During Triassic times, however, those high, grandiose mountains were still far in the future, and the first tectonic compressions only arched and warped the land, producing wide, subsiding basins and depressions that became dotted with lakes and streams. The area now known as Arizona was at that long-ago time on the equator. Unlike its current, arid state, it was a place of wetness, humidity, and jungle vegetation among the many swamps and riverbanks. Great thicknesses of sand and mud were deposited in this region, preserving, in the process, a rich record of the abundant life existing during the latter part of the Triassic Period. The strata of modern-day Arizona are found in the Painted Desert, a colorful panoply of green, blue, gray, and, especially, all hues of the red rocks so characteristic of the Early Mesozoic Era. The best fossils from this rich sedimentary record are found in the Petrified Forest of Arizona, preserved as a national monument since early in this century thanks to the efforts of John Muir.

Fossils recovered from the Petrified Forest and elsewhere in the Painted Desert have given us a rich view of ancient life. Unlike the Connecticut River Valley, where footprints but no bones could at best tantalize and tease the curious paleontologists, the thick rainbowed rocks of Arizona have yielded a trove of fossil bones, shells, and plants. It seems likely that many of the same creatures leaving their footprints in the contemporaneous Connecticut River Valley deposits also lived in the Painted Desert region, for the rocks are of the same age, and the bony feet recovered from the Arizona deposits seem to match the footprints left in the eastern brownstones.

Over 200 species of plants and 60 animal species are known from the

Petrified Forest alone. The forests and swamps were dominated by conifer trees, some of which were at least 200 feet tall; other plants included giant horsetails, not unlike the modern living fossil *Equisetum,* as well as myriad ferns, cycads, and club mosses. Not a flower would have been seen, however, for the evolution of flowering plants was still far in the future.

The assemblage of land vertebrates during the last half of the Triassic Period was vastly different from the paltry assemblage of First Event survivors that started it. *Lystrosaurus* was long gone, but several of its dicynodont descendants still existed, of a size and appearance similar to rhinoceroses. Other protomammals also lived along the riverbanks and shores of ancient Arizona, but never in large numbers; the protomammals were nearing their end, and in far-off South Africa they had already made the crucial evolutionary transition from reptile to mammal. Amphibians, also survivors of the First Event, had themselves made a critical evolutionary transition. In the oldest beds of the Painted Desert are found the skeletons of the largest amphibians ever to have lived, huge, half-ton monsters looking like bloated salamanders. These surely ugly, ungainly creatures, known as metoposaurs, probably preyed on the numerous fishes and smaller amphibians and reptiles living in this dismal swamp. But like the protomammals, they too were but crucibles of evolutionary change, outdated body plans made obsolete by their more modern descendants, the salamanders and frogs that still exist so successfully today. Before the end of the Triassic Period, the metoposaurs were extinct.

If we could go back to the long-ago Arizona during Late Triassic times, our first impression of the scene might be familiarity rather than great strangeness. The riverbanks would be heavily vegetated, but the flora would, on first glance, appear not too dissimilar from tropical scenes of our world. And the numerous, large, crocodilelike reptiles sprawled on the muddy shores and cruising through the sluggish rivers would look quite African, a vision out of any Tarzan movie. Granted, they were pretty healthy specimens for crocodiles, with lengths of fifteen to twenty feet, and some as long as thirty feet, and, under closer scrutiny, it must be admitted that these "crocodiles" are a bit peculiar looking, with nostrils perched far back on skinny snouts, almost at eye level, and yes, about a dozen have crawled off their muddy bank and are sculling this way in quite determined fashion. . . .

And as the great, crocodilelike phytosaurs splash out of the water, yellow eyes fixed on me, my last impression before I sprint into the forest is that they run remarkably well on land, since their legs are underneath them, not sprawled out to the side as in modern crocodiles and alligators. The forest is very thick, and the vegetation around me is slashing my face and arms, but I appear to be putting some distance between myself and the pursuing phytosaurs. Climbing up onto a huge, fallen log, I take stock. Scratched, out of breath, and ruing my circumstances, I now have time to better survey my Triassic surroundings. The jungle is alive with sound, every conceivable octave jammed with chirping, thrumming, croaking, and rasping calls for mates, cries for help, pronouncements of territory, announcements of life. Numerous insects flit about, dragonflies being most common, and high in the arboreal canopy above me I see a flying reptile flapping awkwardly between trees; but there are no birds to be seen, nor are there any flowers or even broad-leafed plants about. The forest is a lush riot of ferns, cycads, pines, and spiky, jointed trees; giant conifers tower over me, so that the forest floor is only occasionally dappled with sunlight. The heat and humidity are stifling. Movement in a nearby glade catches my eye, and I see about a dozen slim animals emerge from the verdant jungle. They are bipedal, with long necks and a thin, straight tail held stiffly out behind them, and about five feet long. As they walk, their heads bob, and I am reminded of giant birds like emus or ostriches in the nervous way they move about. They have two short arms hanging down, and as I watch these graceful creatures I realize that I am looking at early dinosaurs. They are vastly different from the sprawling protomammals in their grace and maneuverability, and I have no doubt that they can attain great speed with their muscular back legs if frightened into flight. With one of their number standing guard, the others begin to paw among the leaf litter of the forest floor in search of food, darting their slim heads downward on occasion to grab a succulent insect or lizard with toothy jaws.

A distant bellow brings the entire herd to attention, all heads pointed in the direction of the noise. And then in a blur of motion they are gone, scattering into the underbrush as a commotion of grunting cries and thrashing bushes heads toward our clearing. A large, armored creature about ten feet long is the first to shuffle into view. Looking something like a giant armadillo, this four-legged tank is an aëtosaur, first of the heavily armored reptiles. Its back is covered with large bony plates, and wicked spikes extend laterally from the sides of its broad back and tail. I assume it is being chased, but as the aëtosaur crosses the clearing and disappears into underbrush on the other side, another creature emerges into view with a predator in close pursuit. The prey, about the same length as the aëtosaur but much less massive, runs on four sprawling legs, its pursuer on two. As they burst into the open I finally get a good look at the two creatures. The first is a type of protomammal called a cynodont, or doglike reptile, so named because of its protruding canine teeth and lupine head; although looking like the mythical devil-hound Cerberus, this fearsome creature is in reality a rather timid herbivore. Its pursuer is anything but meek-looking, however. I had been expecting to see a staurikosaur appear, a five-foot-high carnivorous dinosaur looking like a small version of *Allosaurus* or *Tyrannosaurus rex*. But the creature crashing into view is far taller than five feet; it towers over its lumbering prey. In the thick underbrush the fleeing herbivore had been holding its own, but once in the open the massive, pursuing predator quickly closes the distance between the two and, with a final burst of speed, launches itself onto the back of the cynodont. Squealing like a stuck pig, the cynodont goes down under the massive weight of the carnivorous dinosaur. The momentum of the two intertwined reptiles carries them to the far side of the glade and then into the surrounding copse of ferns, which they beat down with mad thrashing. The carnosaur comes up on top and, holding its struggling prey with the weight of its body, uses huge toothed jaws, armed with daggerlike teeth, to virtually rip the cynodont's head off.

I sit on my log in the silent aftermath, the forest quiet save for the cracking of bones as the victor feeds on its dead prey. I finally get a good look at the carnivorous dinosaur. Two gaudy crests run along the top of its head; I realize that I have seen a skeleton of this creature in the foyer of the Geology Building at the University of California at Berkeley. It is a dilophosaur, supposedly a Jurassic dinosaur, not one from the Triassic (yet here it is!), and also a star player in author Michael Crichton's recent book, *Jurassic Park*. But in that book the dilophosaurs supposedly spit poison and hooted like owls. As the nonspitting, nonhooting carnosaur contentedly works on its lunch, I wonder how Crichton dreamed up all that stuff. Getting ready to leave, I look down and see another member of this Late Triassic fauna, a tiny ratlike creature, also staring, transfixed, at the scene of carnage before us. It is a mammal, the first true mammal, newly evolved in these Late Triassic days. The huge cynodont being devoured before us is a close cousin, with a shared ancestor to the tiny watching mammal, well hidden on this fallen tree. The last protomammals were found in Late Triassic and Early Jurassic times; their descendants, the true mammals, would watch their final passing. And then, for the next 140 million years, the mammals would hide and quake while the true masters of the earth, the dinosaurs, lorded over their hegemony.

7

To me, the Triassic Period was a time characterized by one of the most interesting assemblages of animals and plants ever to have lived together on the earth. It was a crossroads, really, between life's first great fauna and its second. The ancient floras of the Paleozoic coal swamps had, by the Triassic, been replaced by the conifers, cycads, and ferns. In the seas, new types of mollusks and corals had replaced the ancient Paleozoic sea creatures, and a whole suite of large, sea-living reptiles had returned to the saltwater habitats of their ancient ancestors. These included ferocious, fishlike ichthyosaurs and mollusk-eating placodonts, the latter looking and behaving much like large seals. And on land, the vertebrate fauna was composed of a mixture of Paleozoic survivors, such as protomammals, early reptiles, and amphibians, joined by new creatures, such as the crocodilelike phytosaurs, thecodonts, and the dinosaurs themselves. Last, the Late Triassic witnessed evolution of what seemed to be a most minor group, the true mammals. By Late Triassic time this great admixture seemed to show distinct trends; the protomammals were clearly dwindling in numbers, and the newly evolved mammals, rather than challenging the new host of dinosaurs for supremacy of the land, faded into the background as tiny insectivores and tree dwellers. Evolution seemed to be shaking out this diverse assemblage of creatures through competition and natural selection, and we can only wonder what the final results of these processes would have been. We will never know, for about 200 million years ago, mass extinction ravaged the land and sea once again.

The Triassic Period came to a close with a wave of extinctions. Sandwiched between (and far less severe than) the First and Second events, this episode of mass extinction ended the evolutionary histories of many animals and plants. Most of the protomammals disappeared, as did the phytosaurs, all large amphibians, and some of the seagoing reptiles. In the sea, the long geological history of the shelled ammonites nearly came to an end, as only a few species escaped the ravages of this period of mass death.

Earth scientists are puzzled about the causes of this mass extinction, and its study is only just beginning. Geologist Paul Olsen of the Lamont-Doherty Geological Observatory suggests that this episode of extinction may have occurred in 100,000 years or less—perhaps far less. Olsen also points out that a large meteor impact crater, the seventy-mile-wide Manicouagan Crater located in Quebec, is of the same age as the extinctions. His inference: that the wave of extinction rolling across the earth some 200 million years ago was brought about by the lethal aftermath of a four- to five-mile-wide meteor colliding with the earth. In 1991 scientists found evidence perhaps confirming this theory, for evidence of meteor impact was discovered in Italy. Many other scientists take a more conventional view, suggesting that the extinctions at the end of the Triassic Period were brought about by climate change or sudden changes in sea level. Whatever the cause, the aftermath was clear, however: The dinosaurs emerged as the dominant land animals, and during the ensuing Jurassic and Cretaceous periods, they reigned supreme and unchallenged. The Triassic Period set the stage for the Age of Dinosaurs.

Chapter Five

The Age of Dinosaurs

I

The Jurassic and Cretaceous periods comprised the great Age of Dinosaurs. With the extinction of the last phytosaurs, aëtosaurs, and most protomammals at the end of the Triassic Period, about 200 million years ago, no other creatures challenged the dinosaurs for domination of the terrestrial ecosystems. For more than 120 million years thereafter they reigned supreme, undisputed monarchs of the land; during that time the dinosaurs flourished, producing a wealth of size and form. And then, 65 million years ago, they disappeared from the earth forever, perhaps the most famous creatures ever to have gone extinct; their very name is symbolic of old, obsolete, uncompetitive, out of date. Their passing was long unlamented and largely unremarked, for they simply seem to have died out due to uncompetitiveness in a very competitive world. But a great revolution in understanding of dinosaur biology has occurred over the last two decades and seemingly proceeds unabated. We now see dinosaurs as far more wondrous and far more cleverly adapted creatures than earlier scientific views seemed to dictate. But this has created a new problem: If they were so great, why are they so dead?

Old view: Dinosaurs were slow, clumsy, and so stupid that they needed a second brain in their pelvic region just to be able to walk; they were cold-blooded creatures with dull gray hides whose time came and went. They sort of faded away in the face of climate change, lots of volcanoes exploding everywhere, and superior competition from the warm-blooded, egg-eating, all-around nasties of the Late Cretaceous Period, the mammals.

New view: Dinosaurs were fast, graceful, smart wildlife with warm blood, brilliant coloring, and excellent parenting skills whose time went by all too fast. They were so wonderful that they really ought to still exist, and the cause of their extinction is a mystery.

This paradigm shift in our conception of dinosaurs is a relatively recent phenomenon and is perhaps best exemplified by the revolution in dinosaur illustration. For much of the twentieth century, the dean of dinosaur illustrators was Charles Knight, who painted for the American Museum of Natural History. He was a fantastically skilled painter, especially since he had to produce portraits of creatures known only from bony skeletons, and quite often incomplete skeletons at that. So real do his pictures look that they resemble photos more than paintings. But however real they looked, Knight's dinosaurs seemed somewhat ponderous, perhaps because all were dragging enormous, heavy tails behind them. Moreover, none of his dinosaurs ever seem involved in anything more strenuous than grazing or slowly walking. Contrast these old paintings with the best from today's illustrators, by talented artists such as John Gurche, Douglas Henderson, and Bob Bakker (to name a few). The new generation of dinosaur illustrators depicts much different creatures. One noticeable change is in posture of the bipedal forms; the new dinosaurs have backbones parallel to the ground, making them look more like ostriches than kangaroos. A second difference is in the attitude of the tail, which no longer drags on the ground behind the beasts but is held stiffly up in the air to serve as a counterweight to the body. The dull, gray, elephant's skin that adorned Knight's dinosaurs has also been replaced, not only with a rainbow of hues but with the most outlandish display and camouflage patterns as well. Most recently I have detected a distinct fashion trend toward gaudy bull's-eye spots on the frills of ceratopsians. But the most radical change of all can be seen in the poses: The dinosaurs depicted by today's illustrators are veritable acrobats. Jumping, leaping, swimming, and sprinting in any number of gravity-defying positions, and involved in complex feats of social behavior or predation, the "new" dinosaurs have smashed away the old stereotypes. The message comes through clearly: Dinosaurs were complex, social, active animals.

2

The great 1990s hoopla about dinosaurs is a bit misleading. The recent news magazine covers and PBS exposure leaves the impression that dinosaurs are "new" discoveries for a clamoring public, when in reality they have been fashionable for 150 years, and box office for more than 80. Sometimes, late in the evening, one of the great old dinosaur movies will find its way onto TV, watched by true aficionados who, like porno viewers, patiently overlook the stilted human characters, terrible acting, and worse dialog to revel in the short scenes of dinosaur action. Like the changeover in illustrations since the time of Charles Knight, dinosaurs portrayed in the movies give a cultural history of their evolution in the popular psyche. Some of these old dinosaur movies, such as the original *One Million B.C., Journey to the Center of the Earth,* and *The Lost World* (the most recent version), used modern lizards with pasted-on spikes and spines as dinosaur stand-ins, which seemed reasonable, for dinosaurs were long viewed as simply very large lizards. But these Hollywood lizard-dinosaurs, on the whole, were unsatisfactory; very few knowledgeable viewers were going to buy a tarted-up *Iguana* as substitute for a *Tyrannosaurus rex.* Better were the dinosaurs made out of clay. They were probably six inches high, models some poor technician laboriously moved in stop-frame action to produce motion. Because of the great labor and obvious expense, these animated dinosaurs were onscreen for very short periods of time. Nevertheless, some great (and many more not-so-great) scenes still exist.

The undoubted master of this technique was Ray Harryhausen. His greatest creations (in my opinion) probably occurred in *The Valley of Gwangi* (cowboys vs. *Allosaurus*) and *One Million Years B.C.* (the remake, starring Raquel Welch vs. *Allosaurus* and gravity, not the classic original with Victor Mature, which had far superior acting but vastly inferior dinos). Although all of Harryhausen's movies involved the temporally impossible interaction of dinosaurs and humans, the action was, nevertheless, spectacular. Harryhausen himself had a great pedigree to create such films, for he learned his trade from the genius who created the greatest dinosaur movie of all: the original *King Kong*.

Willis O'Brien was the special effects wizard who brought *King Kong* to the screen. I recently had a chance to see an unedited version in a real movie theater, rather than on a tiny TV screen (where I have seen it at least five times before. But I bet that true dinosaur buffs like Bob Bakker and Jack Horner have seen it more times than I). Several minutes of footage from the first half of the film had been restored; long ago, squeamish TV executives had edited out scenes dealing with the rather grisly ends of various crew members who chased after the departing great ape. Happily, they left intact the greatest scene of all, the highly symbolic killing by King Kong ("extinction by mammals") of the best *Tyrannosaurus rex* ever to appear in any film. With the restored footage, and on the big screen, the film is glorious. It is even great in the colorized version.

King Kong probably had as much to do with public awareness and image of dinosaurs as any other medium. But in creating the dinosaurs as he did, O'Brien left a very false image. O'Brien's dinosaurs were universally ferocious (every one of them, including the surely docile herbivores such as *Brontosaurus* and *Stegosaurus* end up attacking the men), and they were all huge and ponderous. But while there were ferocious, huge, and ponderous dinosaurs during the Mesozoic Era, it is doubtful that very many dinosaur species could be characterized as being all three. O'Brien should not be taken to task, however, for his vision of dinosaurs was certainly a product of consultation with paleontological experts of the time. While it is doubtful that many would have advocated having *Brontosaurus* chasing a man up a tree and then snatching him out of it with a great mouthful of sharp, spiky teeth, the public at the time was probably unbothered, for most dinosaurs were depicted as ferocious. It is also probably no accident that this giant sauropod is first seen in the film living in a lake, for early-twentieth-century paleontologists could not conceive of such large beasts walking around on land for extended periods; they felt sauropods probably spent much time in water to support the great mass of their bodies. The posture of the various dinosaurs, with their massive dragging tails, is also surely straight from some professor's mouth. These reconstructions, like the paintings of Charles Knight, have been the favored interpretation of dinosaur posture, gait, and appearance since early in this century.

King Kong was made in the early 1930s, while Harryhausen's *One Million Years B.C.*, with its somewhat similar looking and acting dinosaurs, was made in the late 1960s. Yet it is no accident that the dinosaurs in the latter film are virtual carbon copies of those in the first; during the thirty years between these two films, paleontologists made no discoveries or reinterpretations of dinosaur biology that would require substantive revision of animated dinosaurs. By early in the 1970s, however, our whole picture of dinosaurs began to change. Tragically (the dino buff speaking again), there had not been a major studio, big-screen movie dealing with dinosaurs since *One Million Years B.C.* until Steven Spielberg's adaptation of Michael Crichton's *Jurassic Park* was released in 1993. Given the enormous wave of popularity the dinosaurs are now riding, however, that will surely soon change. And it is a safe bet that like the dinos in *Jurassic Park,* Hollywood's future visions of these Mesozoic icons will be quite different from past films, filled with the same colorful, acrobatic, energetic dinosaurs currently being painted by late-twentieth-century illustrators, with not a dragging tail or water-supported sauropod to be seen.

It is not only the artistic rendering that exemplifies our changing views about dinosaurs, for museum displays are being hurriedly rebuilt to better reflect newly interpreted poses and postures. This trend is perhaps nowhere better seen than at the granddaddy of all dinosaur museums, the American Museum of Natural History, which is in the process of standing its *Barosaurus* up on hind legs, after eighty years of tail-dragging, four-legged gait.

In paintings, movies, and exhibits, dinosaurs have undergone a metamorphosis. What happened?

3

One would think that the sweeping revolution in our understanding of dinosaurs would have been provoked by new discoveries or by the breakthrough research of some budding paleontological genius. Yet while there have been some spectacular new finds, and brave intellectual voyages following them, the driving force behind the dinosaur renaissance was a man who has never published a paper about dinosaurs in his life.

"Revolutions" in science are brought about when we begin to look and think about nature in entirely new ways. In the early 1960s, Professor Steven Wainwright of Duke University began to think and publish about biological structure in a way different from most previous naturalists: He brought the methodology of physics and engineering to biological interpretation.

The interpretation of biological form is referred to as functional morphology; since every form was created through natural selection, it has been widely believed that it must have some (or many) efficient functions. But much of the biological literature about functional morphology has involved "just-so" stories; structures have been examined and then interpreted based on inference, experience, and intuition—human intuition. Another of the favored methods of functional interpretation, especially in the study of extinct animals, has been to use modern analogs as models for interpretation of no longer existing structures. The interpretation of dinosaur form and function has featured all of these methods. The story of *Brachiosaurus* provides an excellent example.

In the early 1900s, a German paleontological expedition excavating in East Africa uncovered the most massive dinosaur yet discovered. Looking something like *Brontosaurus,* but far more massive, the newly christened *Brachiosaurus* must have weighed between 80 and 90 tons, making it by far the largest land animal ever to have lived. (There have been recent discoveries of two sauropod species perhaps even more massive than this. Named *Ultrasaurus* and *Seismosaurus* by their discoverers, these skeletons are still far less complete than the *Brachiosaurus* skeleton discovered in Africa, so their claims for the honor of "world's biggest" are still premature.) *Brachiosaurus* was so big that its discoverers considered it inconceivable that it could have walked around on land. Paleontologists therefore decided that this great creature must have lived its life completely submerged in lakes (hence the submersible *Brontosaurus* in *King Kong*). This interpretation was further strengthened by the discovery of a complete *Brachiosaurus* skull, showing that the nostrils were high *atop* the head, rather than at the front of the head as in most animals. This finding provoked one of the longest running "just-so" stories ever made up about dinosaurs. *Brachiosaurus* was illustrated as living and walking around on deep lake bottoms, with only its nostrils perched out of the water. Its long neck, stretching the tiny head forty-five feet above its feet, was interpreted to act as a long snorkel.

As a boy I found an illustration of a *Brachiosaurus* in this submerged position, with its feet on the bottom of a lake, long neck stretched upward, and nostrils barely breaking the water's surface. This picture, which I taped to my bedroom wall, inspired me to experiment with snorkels. Further, corroborating evidence of the snorkel hypothesis was seemingly found in several episodes of Walt Disney's *Davy Crockett.* Whenever the King of the Wild Frontier (or any number of other movie and TV heroes including James Bond) was threatened by overwhelming odds, he simply broke off a convenient hollow reed and jumped into a river or creek, breathing through the reed while standing upright underwater. I tried this trick on several, near-disastrous occasions. While snorkels work perfectly well when you float on the water's surface, causing your lungs to be at most several inches below the water's surface, they do not work well at all if you are standing on a hard bottom, an orientation that places your lungs more than a foot below the surface. Granted, *Brachiosaurus* was a very big animal, with strong chest muscles. Nevertheless, there is no way that it could ever have pulled air down to lungs located twenty or thirty feet below the surface of any body of water. Pressure increases by nearly fifteen pounds per square inch for every thirty-three feet of water depth; the lungs of our *Brachiosaurus* would have had nearly two atmospheres, or over thirty pounds, of pressure squeezing against every square inch of its rib cage. No animal yet on the earth would be able to draw air down a thirty-three-foot-long pipe.

I was able to convince myself of this through actual experimentation, nearly drowning with snorkels of varying lengths. Steven Wainwright, on the other hand, surely would have solved the problem in more elegant (and painless) fashion, using physics and mathematics.

Wainwright's school of functional morphology, christened biomechanics, routinely applied laws of physics and principles of engineering to a whole spectrum of biological problems. Aided by a group of brilliant students, Wainwright looked at many problems that either had been neglected or had supposedly been "solved" through analogy or "just-so" stories. Topics pursued by Wainwright and his students included the physics of biological flight, the mechanics of swimming and floating, and many analyses of land-animal locomotion. The biomechanics school paid particular attention to aspects of scaling and anatomical design. These studies not only revolutionized the study of living animals, but they changed the way paleontologists looked at extinct species as well. And slowly, case by case, a new generation of paleontologists realized that the old view of dinosaurs needed revision.

Biomechanical analysis was only a tool; to be useful, it needed to be wielded by scientists interested in answering specific questions. But in the heady days when other fields of biology were making great advances by applying rigorous analyses to the study of form and function, paleontology, by and large, languished in its doldrums of required field mapping and obligatory specimen collecting by all its prospective students. The collection of dinosaurs and other vertebrate fossils had put paleontology in the forefront of science during the last part of the nineteenth century and the early part of the twentieth; fifty years later, however, that glamour and vigor had gone. Through a combination of unwitting mistake and bull-headed obstinacy, during the early 1900s paleontology as a discipline became subsumed into geology, rather than zoology or evolutionary biology. Because of this, all prospective students had to become competent field geologists first and competent evolutionists later, if at all. In this system lots of bones were dug up, but their biological interpretation languished. While the fields of genetics and population ecology became the driving forces of evolutionary study during the middle part of this century, paleontology had less and less to contribute. More often than not, a hopeful new paleontologist, arriving in some professor's office for the first time, would be sent away with some assigned geologic quadrangle to map, or some quarry to excavate in the hopes of finding one more new species, or refining some stratal sequence of time. The exciting topics of other fields, such as the ecological interpretation of entire faunas, or evolutionary questions that could be answered only by analyzing already collected material, were usually denied as possible thesis topics.

Invertebrate paleontology began to change in the 1960s with the introduction of computerized studies using vast data arrays. Mathematics, long the forgotten orphan of paleontological study, became a required partner in research efforts. This change came more slowly to vertebrate paleontology. At a few institutions, however, new methods, such as Wainwright's science of biomechanics, were quickly embraced.

One such place was Yale University. There, a brilliant anatomist named John Ostrom had been puzzling over one of the most fascinating yet perplexing evolutionary transitions of all, the origin of birds. Ostrom and many other paleontologists had been intrigued by the similarity between birds, whose first fossils are of Jurassic age, and small dinosaurs. The earliest bird fossils showed greater similarity to certain dinosaur fossils than to those of any other reptiles. Yet birds are active, warm-blooded creatures, while dinosaurs were always considered to have been slow, stupid, cold-blooded ones. The evolutionary transition from dinosaurs to birds, as envisioned by Ostrom, would thus require far more than anatomical change: It would require a complete revamping of the birds' metabolism as well. Ostrom and other paleontologists began to wonder if the reptilian analog long applied to dinosaurs—that, like all living reptiles, they were relatively inactive, cold-blooded creatures—might be incorrect.

A key player in this whole story was a relatively small dinosaur Ostrom discovered in 1964. At first it seemed but another smallish bipedal carnivore, not unlike many other dinosaurs of its time. But further discoveries of this species revealed two anatomical features that surprised Ostrom. The most obvious was a large, wickedly sharp claw on the foot. Ostrom used this feature to give the creature a name: *Deinonychus,* or "terrible claw." The scimitarlike claw was clearly a weapon to wound and bring down prey. But to use it, *Deinonychus* would have had to leap onto its prey, all the while slashing with its foot. Such activity was decidedly undinosaurian, at least according to the traditional view of dinosaurs.

But if the large claw on *Deinonychus* was visible proof that at least some dinosaurs lived an active life, small bones found along its tail were even more important in changing Ostrom's view of dinosaur biology. He found bony rods that must have acted as stiffeners, producing a rigid tail that could not have been dragged on the ground behind the dinosaur. With these bones, Ostrom had found proof that this dinosaur used its tail as a stiff counterbalance behind its body. A picture emerged of a swift, agile, dynamic dinosaur. One of Ostrom's students drew a picture of what *Deinonychus* may have looked like while running. This now-famous picture, by a young man named Robert Bakker, crystallized the image of the "new" dinosaurs as fast, alert creatures; *Deinonychus* and closely related forms such as *Velociraptor* became the symbols of the "new" dinosaurs.

Velociraptor, whose name means "fast-running robber," was discovered by the first American expedition into Mongolia's dinosaur-rich Gobi Desert in 1921–1922. But the discoverers of this surely swift creature overlooked the significance of tail stiffeners and the fact that *Velociraptor* had one of the largest brains known of any dinosaur, as well as legs and pelvis suggestive of a swiftly moving predator. I suspect that packs of *Velociraptor*s—one of the stars (or villains, depending on your perspective) of Crichton's *Jurassic Park*—or *Deinonychus*es would compete quite successfully on our earth today, in places such as the plains of eastern Africa. They may have been similar to prides of lions or packs of wolves, and perhaps they, rather than we, would have risen to become the earth's top carnivore had they lived among and preyed upon the first populations of our own species, a million years ago.

Using biomechanics as a tool, scientists such as Ostrom, Bakker, Kevin Padian, David Norman, and others brought dinosaurs to life as never before. The myth of water-supported sauropods was demolished by structural studies of *Brontosaurus* leg-bone dimensions; the old saw about two-brained yet stupid *Stegosaurus* was debunked by examinations of dinosaurian brain anatomy. Pterosaur flight was studied, and dinosaur walking was analyzed by the study of trackways. But perhaps the most influential and controversial studies of all originally came from Ostrom's *Deinonychus:* In 1968 Ostrom gave an oral presentation in Chicago suggesting that at least some dinosaurs had been warm-blooded.

The warm-blood versus cold-blood controversy dealt with far more than alternative methods of dinosaur metabolism; it was really a battle between the old and new interpretations of dinosaur biology. Championing the cause of "hot-blooded" dinosaurs were Bob Bakker, Armand de Ricqlès, and later, Jack Horner. Following the same arguments later used to suggest that the protomammals were warm-blooded, such as bone histology and predator-prey ratios, Bakker and the others argued that cold-blooded creatures could not have been as successful as the dinosaurs were. Opposition to this theory was great, however. A conference on the subject, resulted in a volume of contributed papers arguing both sides of the debate, which was titled *A Cold Look at Warm-blooded Dinosaurs,* leaving no doubt where the sympathies of the book's editors lay. Ostrom eventually backed away from assertions that all dinosaurs were necessarily warm-blooded, for studies on large reptiles showed that as size increases, the differences between warm-blooded and cold-blooded physiologies disappear. Large dinosaurs, such as *Brontosaurus,* were protected and insulated from ambient temperature changes by their very bulk. But the smaller dinosaurs, such as the fleet carnivorous species, would surely have benefited from warm-blooded metabolisms and, according to many specialists, could not have existed without such a physiology.

The arguments favoring warm-bloodedness in dinosaurs were greatly strengthened by the finding of dinosaur fossils high above the Arctic Circle by William Clemens of Berkeley. Even considering that the Cretaceous world was warmer than now, these high-latitude dinosaurs nevertheless would have had to endure great cold during the long months of winter darkness in these arctic habitats. (Or perhaps they made long seasonal migrations to avoid the arctic night.) The largest reptiles now on the earth, crocodiles, studiously avoid even temperate habitats, let alone cold ones, in favor of the tropics. It is difficult to envisage cold-blooded dinosaurs surviving in the arctic wilderness, even during the long days of the summer.

The crowning studies of the dinosaur renaissance came from paleontologist Jack Horner. His pioneering work in northern and eastern Montana has perhaps brought dinosaurs to life in our minds like no other. Horner's work on dinosaur egg sites and his insights into nesting and parenting behavior put the final nail in the coffin of the old dinosaur mythology. We are left with a vision of complex, well-adapted animals —animals that mysteriously disappeared 65 million years ago in one of the earth's greatest extinctions.

4

The most curious aspect of the dinosaurs' extinction is perhaps not the identity of their killer, but the fact that the very paleontologists who have shown such intellectual bravery in changing our view of dinosaurs act so muddled when it comes to understanding their demise. Of all of the dinosaur specialists I have talked to, only one, Dr. Dale Russell of Toronto, is willing to accept that these great reptiles may have died out suddenly rather than gradually. This may be because the dinosaur guys are right—the extinction of the dinosaurs during the Second Event *was* a long-drawn-out process. But a great deal of evidence suggests that this was not the case, and in a scientific sense the vertebrate paleontologists are becoming increasingly isolated in their insistence on gradual extinction. Even the most strident of all of the Young Turks, Robert Bakker of Colorado, still embraces what may be the oldest of the Dinosaur Heresies: that dinosaurs simply faded away over a protracted period of time. According to Bakker, some terrible disease did in the dinosaurs, and wandering dinosaurs gradually transmitted this lethal germ all over the globe. But from the man who has so successfully challenged the old orthodoxy of dinosaur biology, this idea seems far less compelling than his numerous other assertions; there is no bacterium or virus known on the earth that could cut such a wide swath, among so many different types of animals. Bakker is asking for the equivalent of a disease that could simultaneously kill off lions, wolves, rats, and whales. Nothing is impossible, but diseases usually are very specific to one species, not entirely different groups of animal families. Bakker and the other dinosaur paleontologists all have a stock answer about dinosaur extinction, for they are always asked. But the lot of them seem far more comfortable talking about dinosaur life than dinosaur death.

Perhaps the most persistent argument advocated by those favoring a gradual extinction of the dinosaurs centers on dinosaur diversity during the last 5 million years or so of their long reign. During that period, were the numbers of dinosaur individuals and species increasing, decreasing, or staying the same? Most vertebrate paleontologists seem to say the numbers were gradually dwindling. It would be the easiest thing in the world to test this hypothesis by simply counting up the total number of dinosaur species known worldwide for the last 5 million years of the Cretaceous Period. The answer, it turns out, is anything but straightforward.

5

The beauty of being a nonspecialist is that one can blissfully simplify things, especially if not encumbered by too many facts. With this caveat recorded, I will venture to say that the Jurassic-Cretaceous heyday of the dinosaurs seems divisible into two parts.

I like to think of the first phase as the age of sauropods. Dinosaur faunas from Jurassic and Early Cretaceous rocks are spectacular for many reasons, but especially for their abundance of long-necked behemoths such as *Brontosaurus, Brachiosaurus, Diplodocus,* and their brethren. Many other dinosaurs existed during this period as well, of course. Armored forms, such as stegosaurs, lived among the sauropods, as did ferocious carnivores. But all in all, the long-necked, long-tailed, elephant-legged sauropods seem the most characteristic and certainly the most grandiose components of these ancient days.

The second assemblage was characterized by a bevy of duck-billed and horned dinosaurs, along with a retinue of carnivorous species large and small. The changeover to this second great assemblage took place during the early part of the Cretaceous Period. It seemingly was not the result of some great mass extinction among the sauropods; instead, it came about through a slow and rather gradual change, involving the replacement of the sauropods with two very different types of plant eaters, the duck-bills and the ceratopsians, or horned dinosaurs. Although sauropods still existed (in fact, the youngest dinosaur beds of all still contain the bones of at least one sauropod species, according to Bob Bakker), their diversity and numbers were far lower during the middle and late parts of the Cretaceous Period. The stegosaurs were also replaced, by other armored dinosaurs such as ankylosaurs.

The carnivorous dinosaurs did not undergo a similar changeover. Although individual genera and species changed, the bipedal carnivores looked pretty much the same; although new, midsize models such as *Velociraptor* and *Deinonychus* don't seem to have earlier parallels, the larger carnivorous species such as *Gorgonosaurus, Albertosaurus,* and the greatest of all, *Tyrannosaurus,* appear to be direct descendants of the Jurassic and Early Cretaceous carnivorous forms found among the sauropod assemblages.

The changeover from dinosaur assemblages dominated by herbivores with long necks, seemingly adapted to browsing from trees high overhead, to herbivores that appear to have been ground grazers may have been brought about by a fundamental change among plant floras that occurred during the Late Mesozoic. About 100 million years ago, flowering plants began to proliferate rapidly and began to displace the fern and conifer-dominated plant assemblages. So successful were the newly evolving angiosperms, as flowering plants are called, that they rapidly dominated the plant assemblages. The appearance of lush, ground-hugging bushes and shrubs may have sparked the changeover among the herbivorous dinosaurs. Whatever the cause, dinosaur experts agree that the diversity in number and form among dinosaurs continued to increase throughout the long Age of Dinosaurs.

Geologists have broken geological time into hierarchical, named units. The longest are called eras, which are in turn subdivided into periods, themselves subdivided into ages. The Cretaceous Period, the last large-scale unit of the Mesozoic Era, is thus subdivided into shorter blocks of time—ages—each about 5 to 10 million years long. The debate about dinosaur extinction really boils down to the number of dinosaurs found during the last two ages of the Cretaceous Period. (These ages are formal time units and should not be confused with informal terms such as the Age of Dinosaurs.) Some paleontologists argue that the dinosaurs were dwindling in numbers for tens of millions of years prior to their final demise; this line of reasoning suggests that dinosaurs were doing a slow fade in the face of long-term environmental change, such as global cooling or warming. Other paleontologists argue the opposite, that the dinosaurs were doing very well, thank you, and then were wiped out by some rapid global catastrophe. The majority of paleontologists don't seem to want to take a stand.

The second-to-last age of the Mesozoic Era, called the Campanian Age, began about 80 million years ago and ended about 10 million years later. It is named after the Champagne region of southwestern France, where rocks of this age are particularly well exposed. The last age of the Cretaceous Period, called the Maastrichtian Age (named for a small town in Holland), is only about 5 million years in length. Its end coincides with the end of the Cretaceous Period and the end of the Mesozoic Era, all brought about by the great mass extinction that ended the era, the Second Event. The Campanian Age was a time of high dinosaur diversity and numbers; more species of dinosaurs are known from this time block than any before or after. During the Campanian Age the ceratopsian dinosaurs increased greatly in numbers through the formation of many new species. Duck-billed species proliferated, and the maiasaurs studied by paleontologist Jack Horner were laying their eggs in northern Montana. But the rich record of dinosaurs from this time may be more a function of the numerous localities preserved of this age than a true record of dinosaur diversity; Campanian beds in Montana and Alberta alone have yielded more dinosaur skeletons than any other localities on the earth, and the large numbers of species known worldwide from rocks of this age may be a reflection more of the large numbers collected than a true measure of dinosaur-species richness. In contrast, fewer dinosaurs are known from the succeeding Maastrichtian Age, in part because it is only half as long and in part because far fewer localities of this age are known. These types of sampling problems make any comparison of dinosaur numbers from different time periods exceedingly difficult.

Professor Bill Clemens of the University of California at Berkeley has spent two decades researching the reptiles and mammals found in the last few million years of the Age of Reptiles and the first few million of the Age of Mammals. He notes that Late Cretaceous dinosaurs have been collected from thirty-three sites on the earth. At only three of these sites, however, can Cretaceous-Tertiary boundaries be found. In other words, there are only three known sites in the world where the last few million years of the Mesozoic Era and the first few million of the Cenozoic Era can be found. This is horrible news: The transition from dinosaurs to mammalian-dominated ecosystems can be studied at only three places on the earth?

All three, it turns out, are found in North America. One of these is in Alberta; a second is in the flat plains of eastern Wyoming. The best of all, however, is in the rugged badlands of eastern Montana. There, at a place appropriately named Hell Creek, the last stand of the dinosaurs, the most prominent members of the Mesozoic land communities, is recorded in the rocks.

Chapter Six

Death of the Dinosaurs

I

The sky on this clear night is awash with stars. The Milky Way is a pale shroud bisecting the dark hemisphere; to the east, Pegasus has risen, and below it Taurus is now striding upward into the black vault overhead. The stars scattered across this Montana sky do not twinkle as they do in the cities; the air is too clear here, too clean. It seems ancient and unearthly, like the light I am watching.

I am camped on the shores of the Fort Peck reservoir in easternmost Montana, a large lake formed when the headwaters of the Missouri River were dammed many years ago. Around me lie many miles of tumbled badlands, twisted rock of rainbow colors by day, but now visible only as pale silhouettes against the night sky. The September evening is still and quiet, a noiseless night save for the buzz of late-summer crickets. Most of the land making up the endless expanse of eastern Montana is fenced rangeland, and although it takes little imagination to conjure up the vast buffalo and antelope herds that roamed here but a century ago, the quietness speaks of a new and far less diverse reality.

With a pair of binoculars I scan the starry sky in the region of Sagittarius, where the Milky Way is studded with distant star clusters, nebulae, and dim spiral galaxies. Some of the light from the stars I see began its journey toward the earth as recently as several years ago; some, however, has been traveling through space at its fantastic speed for far longer. Surely some of the faint light studding this Montana sky began its long voyage at the same time that the land around me was a forested river valley, lush and verdant, a place of flowers and warmth, of tepid rain and huge herds of animals grazing on the fertile plain, a time of gentle tropical winds and teeming life, the time of the last dinosaurs.

I turn to look at other dim objects far off in the night sky, whose light is but the fossil remains of once-blazing stars perhaps no longer shining, or changed into objects larger or smaller, brighter or dimmer, or of color many hues different from the brilliant blues and reds and whites sprinkling the sky above me. When we look at the stars we are looking into the past. The infinite stars are each a window back into time, for these bits of starlight are real—as real as the bones buried in the badlands whose dim outlines are barely visible around me. The light from the stars and the fossils I collect have both come down to us across the wasteland of time, forming, at best, misty images of a long ago.

I put my binoculars away and slide into a warm sleeping bag, grateful to escape the ever-colder night air. The days of this trip, spent searching for dinosaur fossils amid the arroyos and buttes of the dry Montana countryside, have been hot; the cooling nights, however, are a reminder that summer is nearly at an end. As I begin to drift toward sleep a brilliant shooting star streaks across half the sky, leaving a luminous afterimage in its wake. Its bright incandescence is due to the burning of a rocky meteor perhaps no larger than a baseball, yet the blazing light caused by this bolide's passage through our atmosphere tells eloquent tales of the energy released by even so small an object. Not for the first time I wonder what it would be like to see the fiery descent and collision of a large meteor bombarding the earth. And as I think of meteors I unconsciously begin to think of the last dinosaurs as well, for the two topics are now inseparably associated in the minds of many paleontologists.

I mourn the passing of the dinosaurs, for who of us would not like to see one of those long-dead beasts, not as the stony bones I uncover, but as roaring reality; who hasn't wondered, in some moment of childhood, why they are all so long dead? Many scientists are now sure they know what felled them: Their killer is thought to have streaked out of the same night sky whirling over my head, a giant meteor smashing into the earth with force and destruction thousands of times greater than the cumulative energy of mankind's nuclear weapons all exploding simultaneously. If this theory is true, the dinosaurs disappeared in agony, in sudden mass death. Surely such a catastrophe would leave an unambiguous signature of authorship. But the event, if it happened at all, occurred 65 million years ago, and the dynamic earth has a way of erasing even the most dramatic history over the long roll of time.

2

The hard, orange sun of early-morning Montana makes a promise of heat for the day; hills barely discernible the night before are now brightly colored: tan, purple, red stripes of sedimentary strata in twisted shapes, shimmering about the surface of the flat lake. From a large bay at one end of the Fort Peck reservoir I begin to follow a creek into the hills: Hell Creek, a name famous in the lore of dinosaurs and dinosaur hunters. The sedimentary strata lining the banks of this small, ephemeral stream are but a tiny outcrop of similarly appearing and aged rocks exposed throughout much of eastern Montana, Wyoming, and parts of the Dakotas. Collectively known as the Hell Creek Formation, these ancient deposits have yielded the youngest known dinosaur fossils in the world—and some of the most spectacular: All of the *Tyrannosaurus rex* skeletons yet uncovered have come from the Hell Creek beds. Discovered more than a century ago, Hell Creek seems a most appropriate name, for the strata along its banks are not only the graveyard of the last dinosaurs but seemingly hold the key to the identity of their killer as well. Tightly embraced within this layer cake of sandstone and shale must be evidence of the catastrophe that killed the dinosaurs.

It is the rocks themselves that tell the story of the end of this world, a tale written on the pages of sedimentary strata making up the chaotic badlands around me. The sedimentary beds on Hell Creek are nearly flat, little changed since their deposition so many millions of years ago. But the creek winds up into the nearby hills, and as you follow it upward you rapidly climb into ever higher strata and thus upward through 10 million years of earth history. The sheets of sedimentary rock I search are slices of time, their countless component sand grains deposited millions of years ago, first at the bottom of an inland sea, then at that sea's edge, and finally on a broad, flat floodplain. All of these deposits were deeply buried as younger sediments eroding off the newly rising Rocky Mountains pushed the Hell Creek strata ever deeper into the earth. After millions of years in their subterranean resting places, the Hell Creek strata were lifted and exhumed to become the landscape of eastern Montana.

The lowest and hence oldest rocks to be seen on Hell Creek are black shales, known as the Bearpaw Formation. Compared to the younger rocks above them, these shales are soft and poorly exposed; they crop out in only a few places. I dig through the crumbly surface layers with my rock hammer, brushing away the weathered chips until fresher rock is exposed. Amid the shale's blackness are harder, whiter objects: lumps of limestone, irregular in shape, which often contain fossil treasure. I find several of these and smash them open with lusty cracks from my rock hammer. The first two are empty, but the third flashes red iridescence in the morning sun as it splits in two. Within this hard nodule rests a fossilized ammonite shell, now exposed to the sun after having been entombed in this lithic casket for nearly 70 million years. I chip away at the surrounding matrix, exposing more of the fossil. A flat spiraled shape emerges, looking somewhat like the shell of the modern-day nautilus, but with chamber walls and ornament far more complex than in any nautiloid. The ammonites, once fantastically successful, are now completely extinct; like the dinosaurs, they too were victims of the Second Event. As the morning progresses I find more of these beautiful fossils and the remains of other sea creatures as well, for these beds were deposited on the bottom of an ocean, far from its shores. It was probably a warm but deep place, far below the surge of waves or clash of storms.

As the sun rises higher in the sky I move up the creek, slowly searching among the low hills on either side, and watch as the nature of the rock changes. The dark shales with their marine fossils are gradually replaced by dusky brown sandstones, and the types of fossils change as well. The ammonites are gone, replaced by countless impressions of snail and clam shells. Some of the thick sandstone beds have eroded into fantastic, contorted shapes, while others seem relatively unchanged since their deposition on an ancient seashore, perhaps 68 million years ago. The top surfaces of many beds are covered with ripples, and as I

PHOTO BY KARIN HOVING.

Hell Creek, Montana, being surveyed by the author.

glance at these familiar-looking, undulating surfaces I unconsciously think of warm sandy seashores of California or of the broad white beaches of my home in Washington State. The illusion of a modern sandy beach is nearly perfect, until you put your hand on these ripples and find an unyielding surface of hard, dead rock instead of the warm playfulness of beach sand.

Climbing ever higher through this record of a long-ago world, I watch as the stratal beds again change color and appearance. The thick brown sands are replaced by sandstones and shales of a dull tan hue;

PHOTO BY THE AUTHOR.

Fossilized sand ripples.

like the underlying beds, these rocks also show ancient sand ripples, but in sizes and complexity far different from those in the rocks beneath. These features are the hallmarks of river deposits, forming not at the edge of a sea but inland, on dry land. The fossils change as well: Most common are the remains of clams, but a type of clam found today only in fresh water. Bits of plant material and broken leaves and twigs are also present, in places appearing as thin black coal beds. But most interesting of all are dark scraps of hard material that show a fine reticulate structure when closely examined. Many of these scraps are tiny fragmented morsels, unidentifiable as to source. But others are larger and have definite shapes; some are from limbs, or vertebrae, and some are teeth and claws of creatures that could only have eaten meat. They are bones: the bones of crocodiles, lizards, small mammals—and dinosaurs.

It is early afternoon when I approach the top of the dinosaur-bearing beds. The bones are not rare; almost each arroyo or hillside shows a fragment or two, and sometimes larger things emerge as well: In 1990 alone, two complete *T. rex* skeletons were found and excavated from the Hell Creek beds. I have not been so lucky on this day. In the rocky landscape the heat is now stifling, and sweat pours off me as I approach the final destination of this day's hike. I am now far up the creek, and the beds over my head are about to change in complexion one last time. The somber, muted tan and brown of the Hell Creek Formation, with its enclosed skeletons of the last Mesozoic creatures, is overlain by brighter, more colorful rocks of red to purple to ocher, all shot through in repetitive display by streaks of blackest coal. These beds are known as the Fort Union Group, and they bear the fossils of mammals, crocodiles, and great amounts of plant life—but no dinosaurs. They were deposited 65 million years ago, at the start of a time unit known as the Tertiary Period, which itself is the oldest unit of time assigned to the Cenozoic Era, our era, the Age of Mammals. In a narrow draw I find the contact between the Hell Creek beds and the overlying Fort Union strata. It lies near a thick coal seam and is exposed as a six-inch-thick layer of gray clay. This thin, innocuous bed is a Cretaceous-Tertiary boundary clay. It marks both the end of an era, the Age of Dinosaurs, and the start of a new one. It also holds the evidence of catastrophe within it: It is like a scar, recording the passage of the knife blade that cut off the long history of dinosaurs.

3

For tens of millions of years a giant seaway bestrode North America, cutting the continent in two. From the arctic regions of the far north to the Gulf of Mexico in the south, this wide yet shallow sea was home to an incredible bestiary of now-extinct creatures. Archaic fish and giant marine reptiles prowled the blue waters in search of prey; monstrous turtles sculled among the wavetops, while on the muddy bottoms giant clams formed shelly pavements in their abundance. Overhead, long-winged pterosaurs and primitive birds circled above the whitecapped waves, occasionally diving and sometimes fighting for surface-dwelling fish. Ammonites giant and small lived and died in countless numbers in the sunlit portions of the sea, their empty shells littering the shallow seabottoms and sandy shorelines. To the west of this sea the Rocky Mountains were rising, pushed skyward by huge volumes of magma welling upward among their roots. In places this magma pooled far underground, slowly solidifying to become giant batholiths of granite, the speckled foundation of any continent. Elsewhere in the rising arc of mountains the magma was less docile, fighting violently to the surface and blasting outward from volcanic cones. These giant volcanoes filled the skies with ash and smoke, creating smokestack pillars of blackness to the west of the inland sea. The shorelines of the seaway were covered periodically with falling ash from the volcanos, creating a rich, fertile soil.

Riotous jungles grew and died in the swampy lowland areas along the margins of the great sea. Rivers both large and small poured into the seaway, carrying into the central sea unnumbered tons of sediment derived from the rapidly eroding mountains. Carried within or buried beneath this settling sediment were the remains of many creatures of that long-ago time: the skeletons of fish, the shells of long-extinct mollusks. And more rarely the rivers disgorged rarer treasures, to fall into the shallow sea or, more commonly, be interred on the floodplains and riverbeds: the carcasses of dinosaurs.

The oldest sediments exposed along the mouth of Hell Creek record the last few millennia of the huge Western Interior Seaway. About 70 million years ago the sea began to flow off the shores, diminishing in size until an isthmus formed between the western and eastern halves of North America, connecting these two separated landmasses for the first time in tens of millions of years. The sea continued to retreat, flowing off the continent to the north and south, until the entire central portion of the continent lay exposed as newly emergent floodplain dotted with shallow lakes. With the disappearance of the sea, climate in the area must surely have changed as well. The Hell Creek Formation was deposited into the river valleys and floodplains of this newly exposed landscape, and its fossil treasures are samples of the animals and plants living there. For 5 million years this newly emerged land, with its rich soil and abundant fresh water, must have supported vast herds of dinosaurs.

The land was alive in this ancient world, this Late Cretaceous world; dinosaurs both large and small covered the terrain. Huge herds of duck-billed dinosaurs grazed on the succulent plants in the fertile river valleys, while vast numbers of horned ceratopsian dinosaurs also browsed on the low vegetation. The largest carnivore of all time, *Tyrannosaurus rex,* may have stalked these herds; or perhaps it was but a giant scavenger, stealing fresh kills made by more agile carnivores. Smaller dinosaurs also existed in this community, such as the coelurosaurs and hypsilophonts, looking much like large, walking birds and perhaps acting not much differently, with bobbing heads and nervous movement, and at least one eye always searching for swift death, surely as common to that long-ago world as it is to ours.

For nearly 5 million years this community lived in its paradise. With the retreat of the sea the climate changed, but not precipitously; rainfall patterns were undoubtedly perturbed, and the land-plant assemblages gradually changed in response. Some paleontologists who have studied this past world feel that the climate changes alone were sufficient to end the dinosaurs' long reign. According to this view, the dinosaurs slowly dwindled in numbers over 100,000 years or more, while legions of newly evolved mammalian species sprouted up around them. But other scientists argue that the sea level, climatic, and vegetational changes recorded from the Hell Creek strata were certainly not sufficient to cause the sudden end of the Mesozoic world, for the long Age of Dinosaurs had weathered countless changes in sea level and fluctuations of climate. According to many respected scientists, a hammerblow from space ended this world, a bolt almost biblical in its starkness. And had that blow not fallen, they say, humans would never have evolved. The dinosaurs held the land in check for millions of years; they were too successful, too abundant, too malleable to the forces of change ever to allow another, lesser group—such as mammals—to gain evolutionary ascendancy.

We, the mammals, apparently inherited the earth not so much because of our vaunted intelligence, or great parenting skills, but our utter meekness and luck. It took a rat-size creature, living frightened in the night, to survive whatever finally killed off the dinosaurs, whether it was colossal asteroidal impact or long-term climate change. But in a strange, ironic way, the death of the dinosaurs is poetic justice (if irretrievably tragic to all seven-year-olds), for our ancestors, the protomammals, were poised to reign supreme on the earth over 250 million years ago, when they were almost extinguished by the First Event. This near demise of all species gave the earth to the dinosaurs. It is fitting that their demise gave it back to us.

4

The mass extinction that killed off the dinosaurs as well as 50 percent or more of the other species then on the earth has been recognized for more than a century and a half as one of the most devastating periods of mass death in the planet's long history. And perhaps because it has been known for so long, there has been no shortage of theories attempting to explain its cause. Some theories have been fanciful, such as world-covering floods; others have been secular, or religious, citing God's will. When I was in grade school the favored reasoning seemed to relate to the perceived superiority of our furry ancestors; during the 1960s and 1970s, some combination of competition from the emerging mammals, in concert with slow climate change, seemed to be the frontrunner in the dinosaur death competition. But over the years other hypotheses have found their way into print as well. Many of these have been crackpot ideas, while others emerge from great good humor. (A tabloid headline blaming Big Game Hunters from Outer Space and Gary Larson's view that cigarette smoking did in the dinosaurs are my two favorites.) But among the crackpot ideas, many reasonable and plausible scenarios leading to the Second Event have come forth. These can be categorized as earthly causes (endogenous) and extraterrestrial causes (exogenous). Among the former are the following:

- The climate grew colder or hotter, with the new temperature extremes killing off various animals and plants. Climate change may have been associated with changing sea level or may have been related to carbon dioxide buildup from volcanism, producing a greenhouse effect, which increased temperatures and killed off organisms in the process.
- Precipitation increased or decreased.
- Great amounts of volcanism covered the skies with black soot and ash, thereby changing climate and killing vegetation.
- Volcanic gases stripped away the ozone layer, causing the dinosaurs (or their food sources) to be killed off by excess ultraviolet radiation.
- New types of plants evolved that were unsuitable for dinosaur nutrition, or even poisoned them.
- Mammals ate up all of the dinosaur eggs.
- Increased stress caused the dinosaurs to produce eggs with ever thinner shells. Eventually no new eggs ever made it to hatching.
- The earth's magnetic field changed poles and, in the process, left the earth exposed to solar wind and other ionizing radiation from space.
- Fresh water spilling out of the Arctic Circle rapidly freshened the world's oceans, killing much marine life and changing climate in the process.

The potential exogenous causes include the following:

- A supernova exploded in the near vicinity of our sun, killing off much life in its aftermath.
- One or more giant comets collided with the earth.
- One or more giant meteors or asteroids collided with the earth.

For most of this century, only the earthly causes have been considered serious candidates, in large part due to a concept known as uniformitarianism, an idea that has held geology in an intellectual straitjacket since the middle part of the nineteenth century. Proposed by the pioneering English geologists James Hutton and Charles Lyell, this principle, stating that modern-day processes are the same as those operating in the past (and can therefore be used to explain the formation of ancient geologic structures), was most useful in explaining away the many unscientific myths of rock and landscape formation then prevalent. Uniformitarianism replaced the concept of catastrophism, which held that the layering of sedimentary rocks and the extinction of past animals and plants were due to a series of world-covering floods. While it was a liberating concept in the nineteenth century, however, the principle of uniformitarianism has become a conservative and constraining scientific force in the twentieth. In its most rigid sense, uniformitarianist doctrine viewed all earth processes as slow and gradual. Because of this, many geologists and evolutionists were unwilling to ascribe the great mass extinctions to any rapid cause or event not currently observable on our earth. Therefore, they favored such normal processes as long-term volcanic activity or climate change as the mechanisms for producing mass extinction. No one had ever seen the effects of a nearby supernova or witnessed the aftermath of a large comet or bolide strike on the earth, and hence these possibilities were largely dismissed. By the 1970s, however, the intellectual climate showed signs of a change.

In 1969, the Paleontological Society, a largely American group of professional paleontologists, listened in what must have been a state of surprise as its outgoing president, Dr. Digby McLaren, proposed that a largely unrecognized mass extinction had exterminated much sea life about 400 million years ago. The shocking aspect of this talk was not the definition of another mass extinction, but rather McLaren's hypothesis as to its cause: He proposed that the extinction was brought about by environmental consequences following the impact of a large meteor on the earth. McLaren subsequently published his address in the *Journal of Paleontology* in 1970. He went straight to the point: The environmental effects following a large bolide impact were sufficient to kill off a significant proportion of the earth's marine life. Although other scientists had already postulated that comets and meteors had often collided with the earth, and that such impacts might even have deleteriously affected past life, no one marshaled the facts and arguments as persuasively as McLaren. However, at that time McLaren had no actual proof of such an impact, such as a suspicious-looking crater of the correct age; he just felt it was the best possibility for the extinction he was documenting.

Although well known among geologists (he was director of the Geological Survey of Canada at the time), McLaren and his message concerning meteors and extinctions received little attention outside of the geological community. Soon afterward, however, one of the greatest physical scientists of all time made a startlingly similar hypothesis.

Harold Urey of the University of Chicago could certainly be categorized as having been a genius, and his Nobel Prize for fundamental contributions to many fields of chemistry was well deserved. But to call him only a chemist would be a disservice, for Urey was a great, interdisciplinary thinker, not intellectually chained to any one scientific field. His seminal work on the origin of life as well as his discovery of the relationship between oxygen isotopes and paleotemperature are two advances that surely would have won him Nobel Prizes in evolution and geology if such awards existed. Therefore, when Harold Urey talked, scientists usually listened. In 1973 Urey proposed that the great mass extinction ending the Mesozoic Era resulted from a comet, the size of Halley's, colliding with the earth. This time, however, the scientific community largely refused to believe Urey's proposal.

Surprisingly, even the support of a scientist of Urey's stature could not galvanize the scientific establishment into accepting or even researching the possibility of cometary or meteor impacts as a general cause of biotic extinctions. The reason was partly due to lack of evidence, for like McLaren, Urey was only making a proposition, not demonstrating evidence of cause and effect. But, as paleontologist David Raup has recently suggested, part of the resistance may also have been related to a conservative, deep-rooted clinging to the principle of uniformitarianism. Although it was well known that the earth was repeatedly hit by large bolides early in its history, as evidenced by the numerous large impact craters of a billion years in age or more, and that numerous, large asteroids are known to cross the earth's orbit every year, few scientists were interested in pursuing the topic.

Harold Urey died soon after his comet-impact hypothesis was published. Had he lived a bit longer, he would have seen his hypothesis become the center of one of the great scientific debates of this century. Digby McLaren was luckier. Today he positively (and justifiably) basks in his prescience of two decades ago, for although many disbelieve, no one now dismisses the notion that impacts from space could have caused mass extinctions in the past. The great change in attitude came from a watershed paper published in 1980.

5

The theory that a huge asteroid smashed into the earth 65 million years ago, killing off the dinosaurs in the process, began, strangely enough, as a study of rock magnetism, not extinction. During the 1960s and 1970s an enormous research effort was devoted to a field called magnetostratigraphy, the science of dating ancient rocks through studies of their magnetic signatures. It had long been known that the earth's magnetic field periodically changes polarity. (Usually it switches every million years or so; over a short time the positive and negative ends of the gigantic magnetic field generated by the spinning earth switch poles.) Under special conditions, the direction of the earth's field at any given time can be imprinted into some rocks. Ancient lava and fine-grained sedimentary rocks such as shale give the most reliable information about magnetic signal, because all lava and many types of stratified rock contain mineral grains that are sensitive to magnetism. When molten lava solidifies into hard rock, or when sediments falling onto a deep ocean floor consolidate, these magnetic minerals orient themselves parallel to the current magnetic field of the earth, and will later indicate if the northern magnetic pole had a positive or negative polarity at the time of the rocks' lithification. If geologists sample a thick pile of lava continuously extruded and solidified over millions of years, they can detect a long history of reversals in the earth's magnetic field. Taken by itself, such information is of little use, for there have been many hundreds of such field reversals over the long roll of geologic time. But if incorporated with other time markers, such as fossils, very precise chronologies recognizable all over the world can be developed.

This type of work is vitally necessary to geology, but dreadfully boring to conduct. Yet one of the wonderful things about science is that you can never really predict where any given project can lead. I am sure that geologist Walter Alvarez, as he patiently sampled a great thickness of limestones near the town of Gubbio in northern Italy in 1973, never imagined that his routine magnetostratigraphic sampling would result in a hypothesis that would rock the scientific world.

Alvarez was working with a large crew of geologists. One, a paleontologist named Dr. Isabella Premoli-Silva of the University of Milan, specialized in tiny fossils called planktonic foraminifera. These microscopic shells came from creatures closely related to amoebas. Unlike the amorphous amoebas commonly found in fresh water, however, the foraminifera produce tiny calcareous shells. They were also subject to rapid bursts of evolution, making them excellent fossils for differentiating different periods of geologic time. By integrating the ranges of these fossils with the magnetic reversals recorded in the same rocks, Alvarez, Premoli-Silva, and the others working on the project hoped to provide a calibration of mid-Mesozoic to mid-Cenozoic time. Within this interval sat the boundary between the two eras.

If the great extinction marking the end of the Mesozoic Era was catastrophic to much life then on land, it was even more so for marine life of the time. Hardest hit of all was the plankton, the vast pasture of single-celled plants and animals floating at the surface of the world's oceans. It had long been known that planktonic foraminifera were particularly devastated by whatever cataclysm caused the great mass extinction, for well over 90 percent of the planktonic species were killed off at that time. Premoli-Silva's patient sampling of the great wall of thick white limestones near Gubbio showed that the extinction horizon could be pinned down to one clay layer.

Alvarez became fascinated by this thin clay stratum, no more than several inches thick. By this time it was already known that the planktonic creatures became extinct during the same magnetic time interval, which indicates that the plankton had died simultaneously all over the earth. Looking at the thin clay layer, Alvarez intuited that the extinction was a rapid event, and not the long, gradual process described in all textbooks. One critical question had to be resolved: How much time did it take to lay down the clay layer?

The imprecision of dating is one of the most frustrating aspects of geology. Resolving narrow intervals of time during eons past is exceedingly difficult or impossible. Although the thin clay layer in Italy was known to be about 65 million years old, no one knew how long its actual deposition took, for modern methods cannot discriminate time intervals of less than about 10,000 years.

Alvarez returned to his home in Berkeley, California, and ran the problem by his father. While most fathers can be depended on for useful advice, this particular father was better than most when it came to scientific problems: Luis Alvarez, Walter's father, was a Nobel Prize–winning physicist, one of the giants of twentieth-century physics. The Alvarezes arrived at a novel approach to solve the enigma of the clay layer. They needed some analytical method of measuring how fast a clay layer accumulated on the bottom of a deep ocean, and they concluded that the concentration of meteoric dust in the sediments could hold the answer.

Meteors constantly fall to the earth. They leave tiny residues of their constituents in environments where relatively undisturbed sediments accumulate, such as on deep, quiet ocean bottoms. As noted in Chapter 3, meteoric material can be discriminated from normal oceanic sediments because meteors commonly contain high concentrations of elements not found in abundance on the earth. Because the Alvarezes knew the rate of accumulation of extraterrestrial material on the earth today, they were confident that, by measuring its amount in the clay layer, they would arrive at a new and accurate estimate for the amount of time it took to produce that critical layer.

As they were both Berkeley professors, Alvarez father and son had access to some of the most sensitive analytical equipment in the world at the nearby Lawrence Livermore Laboratory. There, resident scientists Frank Asaro and Helen Michel first irradiated the Gubbio clay samples using a method called neutron activation analysis. Then they measured the concentrations of iridium and its sister element platinum, two metals vanishingly rare on the surface of the earth but found in high concentration in meteors. To make sure that no sampling or analytical errors entered the work, sediments from both below and above the clay layer were analyzed as well.

The first results were completed in June of 1978. The team was astounded at the numbers produced. The clay layer held thirty times as much cosmic material as did any of the sediments below or above.

The Berkeley team debated for a year before going public with their findings. At first they thought that the anomalously high amounts of metals were caused by the explosion of a supernova in the vicinity of our sun; this hypothesis was soon rejected, however. In 1979 Walter made the first public announcement, and in 1980 the Berkeley group published its findings in *Science* magazine. Their conclusions were like scientific bombshells: Based on the concentrations of platinum group elements found in the Italian clay layer, as well as at two other Cretaceous-Tertiary boundary layers (or KT boundary layers, as they came to be called), one from Denmark and one in New Zealand, the Berkeley group proposed that an asteroid at least six miles in diameter struck the earth 65 million years ago—and that the environmental after-effects of this impact caused the great mass extinction, the Second Event. The ultimate killer, according to the Berkeley group, was a three-month period of darkness, or blackout, as they called it, following the impact. The blackout was due to the great quantities of meteoric and earth material thrown into the atmosphere after the blast. It lasted long enough to kill off much of the plant life then living on the earth, including the plankton. With the death of the plants, disaster and starvation rippled upward through the food chains.

Unlike the similar, earlier pronouncements of McLaren and Urey, the Alvarez impact theory had copious, direct evidence and was, for the most part, testable. If such an impact occurred, there should be evidence of it at many places on the earth as well as a large crater. And most important, the fossil record should show a sudden die-off at the level containing meteoric debris. The scientific world was electrified; the bold theory praised and vilified. Scientists, galvanized into action, set out to prove or disprove the theory, through experiments and observation.

Many things happened very quickly. Independently of the Alvarezes, a paleontologist named Jan Smit of the Free University in Amsterdam also concluded that an asteroidal collision had caused the great end-Mesozoic extinction. In 1981 Smit and a colleague discovered an iridium-rich clay layer in southern Spain that coincided with the extinction of over fifty species of planktonic foraminiferans that had lived in the seas in this region; apparently only one species survived the catastrophe. More exciting, however, was Smit's discovery of tiny spheres of rock within the iridium-rich layer. Less than one millimeter in diameter, the shape and composition of these spherules suggested that they were microtektites, or tiny bits of rock melted and blasted into space by the violence of the asteroidal impact. After circling the earth in low orbit, they gradually settled in the days and months after the impact. In the same year paleobotanist Robert Tschudy showed that the boundary layers could be found in terrestrial as well as sediments of marine origin, when he and other geologists found iridium and plant fossils in rocks from New Mexico. Tschudy also showed that a significant break in the fossil plant record occurred at the same time as the iridium anomaly in marine sediments; the pollen from normal plants found in that region at the time suddenly disappeared, to be replaced by a pollen and spore assemblage made up almost completely of fern material. Ferns are well-known "disaster" species, because they quickly move into and colonize disturbed landscapes, such as newly burned land; the first flora to colonize the forbidding landscape following the May 1981 eruption of Mount St. Helens in Washington State was dominated by ferns. Tschudy called this finding from the Cretaceous-Tertiary boundary rocks the fern spike. It was the strongest evidence found to that date indicating that, on both land and sea, a sudden calamity had struck the earth.

The year 1981 also witnessed a most unusual scientific meeting. High in the Utah mountains, at a ski lodge named Snowbird, 120 scientists gathered to talk and present evidence about the Alvarezes' impact theory. The unusual aspect of this meeting was related to the scientific makeup of the participants. Perhaps for the first time, equal numbers of astronomers, atmospheric scientists, planetary geologists, and paleontologists sat in the same room and listened to one another. This interdisciplinary assemblage had found a common interest.

By 1984 high iridium concentrations had been detected at more than fifty Cretaceous-Tertiary boundary sites worldwide. But by that time opponents of the impact hypothesis also had been marshaling facts and discoveries, and countered with a most critical series of papers. Professors Charles Officer and Charles Drake led the opposition, claiming that all of the physical evidence yet found at the various Cretaceous-Tertiary boundary layers could have been the result of massive volcanism rather than the after-effects of a meteor or cometary impact. Their major piece of evidence came from a study of Hawaiian volcanoes, which were shown in 1983 to emit iridium during eruptions. The iridium found in the KT layers, according to their argument, very easily could have come from terrestrial rather than extraterrestrial sources. And if the iridium came from the earth, the extinctions also would have had an earthly cause. Just as things seemed to be moving in favor of the anti-impactors, however, a new bit of evidence was found that seemingly removed most doubt about a 65-million-year-old meteor impact.

While examining newly discovered Cretaceous-Tertiary boundary clays from the western United States, geologist Bruce Bohor noticed an odd bit of evidence. Rather than simply analyzing the critical clay layer for iridium, Bohor disaggregated the clay using solvents and began looking at the actual mineral grains of the layer itself. He began to notice that some tiny grains of quartz and a few other minerals showed curious, microscopic lines running across their surfaces. Bohor and several other geologists identified these structures as shock lamellae. Shock lamellae are known to occur at only two places on the earth: at meteor impact craters and in craters produced by thermonuclear bomb explosions. Within a year similar shocked quartz grains had been discovered from many Cretaceous-Tertiary boundary sites worldwide.

Perhaps the most disturbing bit of evidence surfaced in 1985, resulting once again from intensive examinations of the boundary clay layers. Wendy Wolbach and her doctoral supervisor, Edward Anders, of the University of Chicago, discovered fine particles of soot disseminated in the clay. The type of soot they found comes only from burning vegetation, and the quantity of soot ultimately found in boundary clays from many parts of the globe suggested that at some time about 65 million years ago, much of the earth's surface was consumed by forest and brushfires. A truly horrifying vision emerged: Soon after the impact, most plants then on the earth appeared to have burned. Some of the fires may have started from the fireball and great heat produced by the impact, but the majority were probably set days later as rocky fragments, initially blown into orbit by the explosive force of the impact, streaked back to the earth as bright fireballs of destruction.

In 1988 another contingent of scientists shuttled off to Snowbird, Utah, to discuss the various new findings. For several clamorous days, author after author presented learned models or research results supporting an impact. Although some dissenters still favored episodes of volcanism to explain the evidence discovered in the Cretaceous-Tertiary boundary clay layers, they seemed a dispirited group. Most of the scientific assembly seemed to be moving toward agreeing that the Alvarez team had been right in the first part of their hypothesis: There did seem almost irrefutable evidence that the earth had been hit by an asteroid some 65 million years ago. All that was missing was the "smoking gun." To this objection, however, the advocates of KT impact had little reply, for their opponents still had at least one valid and unanswered criticism: If such a big rock from space hit the earth, where is the crater?

The absence of a crater of suitable size was indeed a continuing problem for the Alvarez group. Their initial calculations, which resulted in an estimate of the incoming asteroid's size, were derived from the amount of iridium found in the first three KT clay layers discovered. After a decade of further discovery and more refined analysis, it appeared that the incoming asteroid had been, if anything, even *larger* than the six- to seven-mile-diameter body first hypothesized by the Alvarez group. This made the absence of an impact crater of suitable size even more embarrassing. Not that the Alvarez team and its supporters were defenseless on this issue. They needed a crater at least 100 miles in diameter, and, paradoxically, such a large structure embedded in the earth's crust is *harder* to distinguish than a much smaller crater. Sixty-five million years is a long time for erosion and other geological processes to work; the crater could have been present on the earth but simply not discovered yet, because it had been buried by younger rock or perhaps changed beyond recognition by mountain building or other tectonic events. A further possibility was that the crater had literally disappeared—that it had been consumed in a subduction zone. One of the tenets of continental drift and plate tectonics is that new crust is constantly being created along the midocean spreading centers. All this crust must go somewhere, however, or the earth would expand constantly. Since the 1960s geologists have known that the "where" is in structures called subduction zones, places where ocean crust slowly submerges into long trenches usually found along the edges of continents. The huge slabs of oceanic crust eventually sink hundreds of miles beneath the surface, where they melt into molten rock. About 20 percent of the ocean basin has been destroyed in this fashion during the last 65 million years. It was thus considered possible that the meteor had struck in the ocean and that its giant crater (sitting in oceanic crust) had been sucked down a subduction zone and destroyed.

While there was no crater large enough to be from the meteor envisioned by the Alvarez group, there was indeed a crater of the correct age, if not size. Many respected earth scientists seized on this structure to support a new hybrid theory: The extinctions had been caused by a combination of volcanic activity and the effects of a small asteroid impact.

Deep under the cornfields of Iowa, near a small town called Manson, lies a twenty-mile-wide crater. It is not visible on the surface, and was long ago buried by new strata and alluvium. Wells drilled into this structure, which was originally discovered by petroleum geologists, returned rock bits dated as being 65 million years old—exactly the age of the Cretaceous-Tertiary boundary and the mass extinction. It was this impact, insisted many Alvarez theory detractors, that resulted in the much-ballyhooed iridium concentrations, spherules, soot, and shocked quartz discoveries. And there the 1988 Snowbird II meeting would have rested, but for the exhortations of a very determined young scientist.

I was invited to the first Snowbird conference, but at the time had declined. I was also invited to the second Snowbird gathering, but this time was required to attend—through the very determined arm-twisting by one of my mentors, I had agreed to serve as coeditor of the proceedings. I attended every talk, discussion, and conference, and with my other coeditor, Buck Sharpton of the Lunar and Planetary Institute in Houston, even had the questions and comments following every formal presentation tape-recorded. Buck and I were chosen to edit the conference proceedings presumably because we were viewed as neutral figures in the debate. We soon found, however, that we were the only ones at the conference who had not taken sides. Each scientist was given fifteen minutes to present findings, followed by five minutes of discussion afterward. Many controversial presentations were made, and much of the discussion was, to say the least, heated. By this time, eight years after the Alvarez group's initial publication, there was not only polarization among the various factions for and against, but there was downright bad blood between many individuals espousing different sides of the controversy. Perhaps because of this, the five-minute question-and-answer period became the forum for much bile-slinging.

Two figures were constantly on their feet following each talk. One, Ed Anders of Chicago, is a chemist who had helped bring to birth the hypothesis of global wildfires following the impact; he showed himself to be enormously knowledgeable not only about the chemical clues to the past catastrophe but in all manner of geological evidence as well. The other was a graduate student from Arizona named Alan Hildebrand. Like Anders, he was pro-impact. But unlike Anders, who attacked what he saw to be flaws in reasoning or logic, Hildebrand was on a crusade. Like a medieval European knight, Hildebrand had come to Snowbird to slay the disbelieving infidels who advocated any cause other than meteor impact. His weapons were a breathtaking mastery of evidence for the event, acquired both from field observation and library research, as well as an equally breathtaking conviction in his belief. I never met a man with less self-doubt, or less humor. But among the great scientific warriors assembled at Snowbird, Hildebrand's unwavering confidence would have been but an irrelevant affectation had he not arrived with a message as well; Alan Hildebrand stood before the 300 scientists and pointed to the Caribbean Sea on a projected map: "There," he said, "lies the long-searched-for crater." And he seems to have been right.

Alan Hildebrand and his academic adviser at the time, William Boynton of the University of Arizona, were convinced that a large meteor had hit in the Caribbean Sea. They arrived at this conclusion through several interdisciplinary routes. First, the mineral assemblages found in various KT clays found across the globe gave important clues. The mineral composition discovered in North American boundary clay layers could have formed only if the meteor had crashed on land; other sites, however, gave equally compelling evidence that it had crashed into rocks that lay beneath the sea. The two Arizona scientists concluded that two impacts must have occurred. One resulted in the Manson crater, but the second site had not yet been positively identified. A second clue came from some very peculiar sedimentary deposits found on a sleepy, snake-infested river in Texas. Paleontologists had long known that uppermost Cretaceous and lowermost Tertiary rocks were exposed along the riverbeds of the Brazos River in central Texas. But unlike many other sites, where the boundary was expressed as a thin clay layer, the boundary beds on the Brazos River seemed to occur in the middle of a thick, rippled sandstone unit. These beds had long been interpreted as having been deposited during a giant storm. In the mid-1980s, however, paleontologists Thor Hansen of Western Washington University and Erle Kauffman of the University of Colorado wondered if the chaotic Brazos beds might not have been formed by a very different event—a tsunami, or tidal wave. Hansen invited a specialist on wave deposits to visit the outcrop. (Good scientists, like good doctors, often ask for second opinions.) Jody Bourgeois, from the University of Washington, was invited and flew down to Texas with Hansen. The beds she saw looked nothing like storm deposits. She made measurements on the ripple marks and other formations and, with the collaboration of Hansen and mathematician Patricia Wiberg, also from the University of

Washington, published in 1988 a stunning paper in *Science* in which they stated that the Brazos River beds appeared to have been made by a wave initially at least 1,000 feet high. No giant earthquake could ever have produced such a huge tidal wave. Bourgeois and her coauthors concluded that the wave was produced by a 65-million-year-old meteor impact.

At this point things began to get a bit muddled about who did what. Hildebrand also visited the Brazos River beds and came to a similar conclusion. Convinced that the impact producing such giant tidal waves had to be situated somewhere in the Caribbean, he began to look for large circular structures by sifting through maps and old geological reports of the region.

At the time of the Snowbird conference, Hildebrand thought that the site may have been off Colombia. Soon after, however, he found information that convinced him to look in a new area: the Yucatán Peninsula. In 1990 Hildebrand and Boynton announced that they had found a large circular structure on the northern Yucatán Peninsula. They interpreted this giant, buried structure to be an impact crater at least 100 miles in diameter. Like the Manson Crater, this newly christened Chicxalub Crater had long ago been buried by younger sediments; unlike the Manson Crater, however, drill cores yielding the most critical information of all, its age, could not be located immediately, having been lost in an oil field fire some years ago. Nevertheless, the coauthors suggested that this crater was indeed the long-searched-for "smoking gun," the site of a meteor impact large enough to have produced a major mass extinction. The incoming meteor had apparently split in two prior to impact, with the larger piece landing in a shallow, limestone-rich sea in the Yucatán, the other hitting the North American continent in what is now Iowa. In an article published in *Natural History* magazine in 1991, Hildebrand credited the discovery of impact wave deposits, such as those found at Brazos River, as being the critical clue leading to the discovery of the impact site. Nowhere did he mention the precise work of Jody Bourgeois and her colleagues, who published the first information about the impact wave deposits.

Several months later a letter was published in *Natural History* in reply to Hildebrand's article. Ironically, it was not from the University of Washington researchers, but from a former Mexican oil company geologist named Glen Penfield. Penfield had spent many years working in Mexico searching for oil. Much of his research involved the use of both gravity and magnetic surveys flown by aircraft. Using this information, he found a large, circular rock body in the Yucatán region that gave gravitational and magnetic signals quite different from those of the surrounding rock. In 1978 he identified this structure as an impact crater, and in 1981 he announced to the world not only that a large impact crater was present in the Yucatán, but that it was probably the same crater the Alvarez group was searching for.

In an astounding case of scientific oversight, almost all those interested in the Alvarez theory overlooked or missed Penfield's hypothesis. In his 1991 letter to *Natural History,* Penfield wrote that he had "identified the Yucatán feature as a probable impact at or near the K-T boundary in the report 'Preliminary Geophysical Interpretation Report —Progreso area, written for Petroleos Mexicanos' " in 1978. In 1981, at the annual meeting of the Society of Exploration Geophysicists, held that year in Los Angeles, Penfield and his colleague Antonio Camargo gave an oral presentation concluding with the following statement: "We would like to note the proximity of the [Yucatán] feature in time to the hypothetical Cretaceous-Tertiary boundary event responsible for the emplacement of iridium-enriched clays on a global scale and invite investigation of this feature in light of the meteoric climatic alteration hypothesis of the late Cretaceous Extinction." Penfield and Camargo had hit a scientific home run: They had found the crater and correctly linked it with the then newly announced iridium findings of the Alvarez group. The problem was, they gave their presentation to other oil company geologists (whose only interest was whether the structure contained oil), rather than university professors, and the information never made it to the Alvarez group. After this public announcement Penfield and Camargo moved on to other endeavors, for they had more important things to do—such as keeping their jobs—than promoting their discovery.

Hildebrand's brief reply to Penfield's letter indicated that he was well versed in the oil company discoveries of the 1970s. Hildebrand deserves great credit for his bulldog determination in bringing to light a long-overlooked discovery; Penfield, after all, had had a decade to make his views and discoveries known. But in amassing and publicizing his information, Hildebrand has been as much bulldozer as bulldog, and those run over in the process will not soon forget.

Further evidence was made in 1990 and 1991. If a giant asteroid had indeed crashed into the Yucatán region, effects of the impact should have been most pronounced in the Caribbean region. This was confirmed when large pieces of glass were discovered by Alan Hildebrand in a Cretaceous-Tertiary boundary site in Haiti. The bed was extraordinary. Where most of the KT clay layers are no more than several inches thick, this bed was several feet thick; where the glassy spherules found in the clay layers were composed of particles at most the size of sand grains, the sediments within the Haiti beds were the size of pea gravel. Analyses of these fragments yielded a date of 65 million years old—the same age as the other Cretaceous-Tertiary boundaries. Soon after, similar beds were found elsewhere around the Caribbean region. The evidence was overwhelming: 65 million years ago, the Yucatán Peninsula was Ground Zero. In 1993, the crater itself was shown by Buck Sharpton and others to be 180 miles in diameter—far larger than previously believed. The KT crater in the Yucatán is now known to be the largest impact crater on the face of the earth.

Late in 1990, a decade after their initial publication outlining the impact theory, Walter Alvarez and Frank Asaro of Berkeley summarized their current views about the topic for *Scientific American* magazine. Ten years of intensive research by hundreds of scientists has resulted in a far more comprehensive vision of the impact and its aftermath. Alvarez and Asaro summarized this new view as follows:

An asteroid or comet at least six miles in diameter enters the earth's atmosphere traveling at a rate of about 25,000 miles an hour. At this velocity it rams a hole through the atmosphere and smashes into the earth's crust. Upon impact, the kinetic energy of the falling body is converted into heat, creating a nonnuclear explosion at least 10,000 times as strong as the explosion that would result from mankind's total nuclear arsenal detonating simultaneously. The crater produced is as large as the state of New Hampshire. Rock in the target area, as well as the entire mass of the meteor itself, are blasted upward through the meteor's atmospheric entry hole. Some goes into earth orbit, while the heavier material reenters the atmosphere after a suborbital flight and impacts the earth as blazing fireballs that ignite the forests. Over half the earth's vegetation burns in the weeks following the impact. A giant fireball also expands upward and laterally, carrying with it additional rock material, which enters and obscures the atmosphere, transported globally by stratospheric winds. The enormous quantity of rock and dust exploded into orbit begins sifting back to the earth over a period of weeks. This material adds to the great dust plume and smoke from burning forests to create an earth-covering pall of darkness. For several months no sunlight reaches the earth's surface; the atmosphere might resemble the oil-fueled miasma covering Kuwait following the Gulf War, but darker. The darkness causes temperatures to drop precipitously over much of the earth, creating a profound winter in a previously tropical world.

The impact creates great heat both on land and in the atmosphere. The shock-heating of the atmosphere is sufficient to cause atmospheric oxygen and nitrogen to combine into gaseous nitrous oxide; this gas then changes to nitric acid when combined with rain. The most prodigious and concentrated acid rain in the history of the earth begins to fall on land and sea, and continues until the upper 300 feet of the world's oceans are of sufficient acidity to dissolve calcareous shell material.

The impact also creates shock waves spreading outward from the festering hole in the earth's crust. Monstrous tidal waves spread outward, eventually washing ashore along continental shorelines, leaving a trail of destruction in their wake.

Following months of darkness, the earth's skies finally begin to clear, the forests to cease burning. The impact winter comes to an end, but temperatures continue to climb even after normal seasonal temperatures are attained, for the impact has released enormous volumes of water vapor and carbon dioxide into the atmosphere, creating an intense episode of greenhouse warming. Climate patterns alter quickly, unpredictably, and radically around the globe as the ever-changing heat balance of the earth—a critical factor affecting climate—attempts to regain some normal equilibrium.

Surely the animals and plants then living on the earth would have noticed all this hubbub.

6

Like most great science, the Alvarez theory is conceptually simple. It is composed of two parts: 1) 65 million years ago, the earth was hit by a giant meteor; 2) environmental consequences of that impact created a great mass extinction. Part 1 seems indisputable. Part 2, however, has been much more difficult to prove. Perhaps a great meteor struck 65 million years ago but had no effect on the earth's biota. This uncertainty about the impact's effects on the earth's Cretaceous organisms caused physicists and geologists studying impact clays and making their impressive discoveries no end of frustration and bad feelings. The subject of their ire was the community of paleontologists, who just couldn't seem to find enough bodies in the right places to match the catastrophic effects envisioned by those advocating giant meteor impacts.

In 1980, when the Alvarez team published their initial report in *Science,* they were convinced that additional geological research would reveal iridium concentrations in all KT boundary layers, a conviction quickly substantiated. They also believed that detailed paleontological investigations of the same stratigraphic sections would show sudden, catastrophic extinctions to occur at the level of the boundary layers. They expected that the last occurrence of all of the prominent victims—dinosaurs, pterosaurs, mosasaurs, ammonites, large clams, and various plankton—would show simultaneous disappearances at the various KT boundaries preserved around the world. The Alvarez group had good reason to surmise this, since the first section they examined, the thick limestones at Gubbio, Italy, showed just this relationship: The fossils of planktonic foraminifera dropped from about fifty species to less than five at the base of the iridium-rich clay layer. It was with mounting disbelief, however, that the Alvarez group found themselves assaulted by a howl of dissent from the world's paleontologists. And the loudest disbelievers of all were those paleontologists who study dinosaurs.

One of the beauties of science is that it is a philosophy of inquiry and thus often seems (to those on the outside) to rise above the plane of ordinary human affairs. But scientific investigation is conducted by people, not dispassionate machines, and all too often emotion and personal beliefs can cloud the purer motives of discovery. Scientists are people, for better and for worse. It is safe to say that both the better and the worse were present in large quantities during the decade-long debate between the impacters and the majority of paleontologists. Sadly, many of the debaters ultimately ended up in mud-slinging contests.

Part of the problem stemmed in no small part from the perceived status of various branches of science. Many physicists view their field as the top of the scientific pyramid, the calling of the best and brightest minds of all humanity, and who could argue with this? In many cases physicists have indeed numbered among their own some of the great thinkers of all time. But some physicists have been barely able to hide their scorn of other sciences, among them Luis Alvarez.

Through a strange accident of fate, the leading paleontological critic of the Alvarez theory worked in the same building at Berkeley as Walter Alvarez. Bill Clemens had spent his career studying the latest mammals of the Cretaceous Period and earliest mammals of the succeeding Tertiary Period. He had conducted much of his field work in the Hell Creek beds of Montana, in an area where the transition between the last Cretaceous and earliest Tertiary beds was thought to be the most complete and continuous of any terrestrial sediments on the earth. If the Alvarez theory was true, he should have seen many dinosaur species becoming suddenly extinct right at the boundary. However, in his many years of collecting, Clemens had never seen a sudden dinosaur die-off; instead, his experience suggested a gradual disappearance of the dinosaurs. In the early 1980s, Clemens again journeyed to Montana and sampled from localities scattered across the Hell Creek basin, at stratigraphic positions where he thought the boundary to be; later analyses of these samples revealed enhanced iridium anomalies at four different sites. Clemens noted that each of these boundary sites was situated just beneath a prominent coal layer, which became known as the "Z" Coal. This particular coal bed could be traced across many hundreds of square miles in Montana and hence became an informal marker bed used to find and delineate the KT boundary. After reviewing past records of his own and those of his students, and making new field collections as well, Clemens concluded that the dinosaurs in the Hell Creek region had been completely extinct well before the deposition of the iridium layer, for no dinosaur material could be found in any of these strata within three vertical feet of the boundary. Because sedimentation was fairly slow in the Hell Creek area, this finding suggested that the dinosaurs were already extinct many thousands of years before the meteor impact. Other scientists studying in the area made somewhat similar findings. For instance, noted paleobotanist Leo Hickey of Yale University suggested that the dinosaur extinctions and the pollen perturbation that came to be known as the fern spike were separated by many thousands of years. In a paper published in 1982 by the Geological Society of America he concluded: "moderate levels of extinction and diversity change in the land flora, together with the nonsynchroneity of the plant and dinosaur extinctions, contradict hypotheses that a catastrophe caused terrestrial extinctions." In staking out these positions, however, Clemens and other paleontologists who doubted impact-caused extinction unknowingly put themselves into the ring with Nobel Prize winner Luis Alvarez, an undefeated heavyweight of science.

By the time of the first Snowbird conference in 1981, the divergence in progress between those scientists explaining the various geological attributes of the boundary clay layers and those studying the fossils beneath these clay layers was already apparent. Rather than confirming the Alvarez hypothesis, most fossil record interpretations seemed to suggest either that major groups went extinct prior to the boundary or were already in decline at the time. Many scientists began to believe that *yes,* perhaps a meteor had hit the earth, but *no,* its impact had nothing to do with the extinctions occurring about that time. Luis Alvarez would have none of this. His tactic was to simply denigrate the methodology of paleontology and personally insult those paleontologists most critical of his theory. Late in his life, in an unfortunate newspaper article published in *The New York Times,* Alvarez took his attack beyond all bounds of scientific decency, calling Clemens "generally incompetent" and others far worse. If it was a heady time to be a scientist involved in all of this, it was also a troubling time.

In retrospect, it is not difficult to pinpoint the reasons for the very different pace at which the two parts of the Alvarez theory were tested and confirmed. Evidence for impact came only from the boundary clays; being only several inches to at most several feet thick, they could be sampled in a day or two and the laboratory analyses finished in several months. The fossil record was a very different ball game. The Alvarez theory requires that various species disappear at the KT boundary. But to demonstrate this, tens to hundreds of yards of stratigraphic section need to be laboriously sampled and collected for fossils. Even worse, the size and abundance of the fossils being sampled play a large part in the story. The microscopic foraminiferans found in the Italian Gubbio section could be studied rapidly because of their small size and great abundance; even small chips of limestone could contain hundreds or thousands of specimens. Larger fossils, however, were different. Dinosaurs, because of their great size and rarity, became the most difficult test of all.

The decade of debate between 1980 and 1990 vastly improved the discipline of paleontology. The healthy cross-fertilization between paleontologists and physical scientists such as astronomers, physicists, and chemists surely improved all fields concerned. Paleontologists began to apply rigorous statistical techniques to their fossil range charts, and they slowly began to appreciate that the positions of fossils in their enclosing strata might provide misleading clues as to whether a given extinction had occurred suddenly or gradually. In my own work on the Second Event I was to learn this lesson all too well.

During the late 1970s and the early 1980s I held a faculty position at the University of California at Davis. Being only sixty miles from the Berkeley campus, we at Davis were quite caught up in the big doings going on there. In 1981 I was invited to Berkeley to give a seminar, for I had just finished a theoretical study showing that one of the prominent victims of the great mass extinction at the end of the Cretaceous, the ammonites, must have gone extinct suddenly for extraordinary and catastrophic reasons. Many of those involved in the impact debate were present, including Luis and Walter Alvarez, Bill Clemens, and David Jablonski. The lecture room was divided into three camps: pro-impacters on one side, anti-impacters on the other, and a confused group in the middle. After my talk the Alvarezes warmly congratulated me for helping prove their point, and they took me to Walter's house for dinner. Bill Clemens left immediately after my talk without comment.

A year later I was invited back to Berkeley. By this time I had completed an extensive collecting trip to Spain to study the actual ammonite extinction pattern in nature, rather than on a computer screen; my findings, which contradicted my earlier work, suggested that the ammonites, like the dinosaurs, appeared to have been completely extinct prior to the deposition of the iridium-rich clay layers in the section I was studying. After presenting these findings to almost the same Berkeley audience I had addressed the year before (Luis skipped this talk, knowing in advance that I would be saying things he didn't want to hear), I was invited out to dinner by Bill Clemens.

I returned to the Spanish stratigraphic sections repeatedly during the decade, collecting more and more ammonites each year. Although many individuals and species could be found tens of yards below the KT boundary, the numbers gradually dropped off as the boundary was approached. After much work I was able to recover ammonites within inches of, but never at the KT boundary. A literal interpretation of my data suggested that a long, gradual extinction of ammonites had taken place prior to the time when the iridium-, spherule-, shocked quartz–rich clay layer had been formed in this region. In actuality, however, the ammonites were simply showing a pattern that came to be known as the Signor-Lipps effect.

While I was pondering the fate of the ammonites, two of my colleagues at Davis, Phil Signor and Jere Lipps, were also considering extinctions and how a sudden, catastrophic extinction might appear among creatures that were relatively rare. Their theoretical findings, published in 1982, suggested that sudden, catastrophic extinctions would appear to have been gradual in any group of fossils unless they were fantastically abundant. The ammonites I collected, as well as many other groups of fossils (including dinosaurs), seem to obey this rule. Paleontologists were much sobered to learn that sudden extinctions would *always* look gradual unless sample sizes and fossil abundance were extremely high.

During the early part of the 1980s the paleontological community was also forced to accept the opposite case as well: Gradual extinctions can look sudden. Many studies of rock accumulation showed that sedimentation is rarely continuous. Breaks in sedimentation are often well marked in the sedimentary record. In fine shales and mudstones, however, the accumulation of sediments can cease for long periods of time, then recommence, without leaving a trace. In such cases many fossils disappear abruptly at one of these discontinuities, looking exactly like a sudden, catastrophic extinction, when in reality the last occurrence terminations were artificially produced by sedimentological rather than biological properties.

By the second Snowbird symposium, in 1988, many new paleontological studies, most based on multiyear field collecting, had drastically altered the view of the Second Event. In sampling techniques and stratigraphic analyses, these studies were far more sophisticated than pre-1980 efforts. Their results were quite clear: More species went extinct at the KT boundary than in any million-year period before or after. But the studies also showed that environmental and species changes were far more complex than would be expected for a single, catastrophic perturbation of the environment caused by a meteor hit. A graduate student from Yale, Kirk Johnson, reported on a magnificent study he had conducted in conjunction with his adviser, Leo Hickey, showing that North American plant species collected from the Hell Creek formation underwent a major pulse of extinction right at the boundary. This study was all the more significant in that it was coauthored by Hickey, who in 1982 had stated that the fossil plant record did *not* support a catastrophic extinction at the KT boundary. But the study also gave clear evidence of complex floral changes at least a million years before the boundary; Johnson and Hickey and the other authors concluded that a catastrophic extinction did occur 65 million years ago, but that it followed and may have been independent of significant climate changes in the region. Two other studies on plant community changes in North America arrived at similar results: A major climate change, followed by an intense and rapid pulse of extinction, seemed to have affected most of North America near the end of the Age of Dinosaurs. One study even suggested an ominous possibility: Minor extinctions immediately prior to the KT boundary all occurred in plants with complex pollen grains; author Arthur Sweet and his colleagues from the Geological Survey of Canada suggested that a sudden loss of insects responsible for pollination may have resulted in the extinction of many plants 65 million years ago. There is very good evidence that many insects and frogs of our present-day world are also endangered and are bearing the brunt of the current extinction, the Third Event; too often, when I point this fact out during public lectures, I am asked the good of having insects anyway. If all of the bees currently on the earth were to disappear tomorrow, much of humanity would soon be very hungry.

Studies from Late Mesozoic oceanic sediments also resulted in a similar pattern of a large, catastrophic extinction following lesser precursor events. It was clear that a hideous, short interval of mass death occurred in the earth's oceans 65 million years ago, wiping out much of the plankton and many larger creatures, such as the ammonites, many other mollusks, and large marine reptiles such as the mosasaurs. These extinctions were surely caused by the effects of the great meteoric impact. But it seemed equally clear that major changes in marine ecosystems had been occurring for 1 to 2 million years *prior to* the impact. At the same time that rapid climate change was affecting the Hell Creek plant communities about 67 to 66 million years ago, great changes were taking place in the sea as well. Water temperatures were dropping, and the amount of dissolved oxygen in bottom waters was increasing. A giant drop in oceanic sea level occurred at this time also. All of these events combined to initiate the Second Event.

My decade-long study of Late Mesozoic strata in Spain and France ultimately showed that the ammonites had remained abundant and diverse right up until the end of the era; the last ammonites in this region were recovered just beneath KT boundary clay layers. Other groups of mollusks recovered from these strata, however, had clearly succumbed long before the ammonites, including two groups that had been among the most abundant creatures on the earth for millions of years. One of these groups was represented by perhaps the most peculiar clams that had ever lived. Called rudists, they resembled miniature garbage cans, with a long, horn-shaped lower shell and a tiny upper shell looking like a garbage-can lid. Although faintly ridiculous in appearance, these clams had achieved ecological greatness, for during the Cretaceous Period they had wrested control of the reefs away from corals.

Reefs have been around for more than 500 million years, and during most of that long interval, the dominant creatures building and holding reef communities together have been corals. During the Cretaceous Period, however, the rudistid clams took over the reef habitats. They grew much faster than corals and took all of the favored places in reef communities. Reefs made up of clams rather than corals would surely have looked extraordinary; surrounded with reef fish and other familiars of today's reefs, clam-constructed patch and barrier reefs and atolls would have been most curious to see. For nearly 40 million years these creatures held preeminence in what is today among the most diverse and favored of all marine habitats, places where sunlight and food are ever abundant. And then, about 1 to 2 million years before the meteor hit, these rudistid reefs began to disappear all over the earth. By the time of the Cretaceous-Tertiary disaster, they existed as only a handful of species in a few shallow-water habitats. The impact totally exterminated these last survivors.

Offshore of the rudistid reefs, another group of bivalves held sway. They were called inoceramids, and it is difficult to convey how successful they were for over 100 million years. We find them in virtually every type of marine sediment deposited during the Jurassic and Cretaceous periods; they seemed to have thrived in all oceanic environments, ranging from shallow lagoons to the deepest seas; one of the surprising results of the U.S. National Science Foundations's Deep Sea Drilling Program was the discovery that giant inoceramid clams lived in water depths as deep as 5,000 feet during the latter parts of the Cretaceous Period; no clams of such size live there today. Although some were small, other inoceramids were the largest clams to have ever lived; specimens over five feet long have been recovered, and individuals a yard long are extremely common. No clams of this size live in today's oceans; only *Tridacna,* the giant clam, comes close, and tridacnas are restricted to a very narrow suite of environments in the tropics.

During my work in Spain and France, it became apparent that the inoceramids, which in some places formed virtual shell pavements across the rocky strata because of their abundance, went extinct about 1 to 2 million years prior to the deposition of the KT clay layer—in fact, it appears that their extinction may have coincided with that of the rudistid clam reefs. For five years I had a brilliant Ph.D. student named Ken MacLeod working with me, analyzing this problem. He showed that the inoceramids disappeared all over the world during a very short period of time, somewhere between 50,000 and 100,000 years, perhaps, and that this extinction episode predated the end of the Cretaceous Period by at least a million years. The extinction of these two important groups of marine creatures had nothing to do with the impact, which occurred much later. It does, however, tell us much about the world just prior to impact. As seen in the plant fossil record on land and among mollusks in the sea, the earth of 67 to 66 million years ago was going through a profound environmental reorganization.

The cause of these pronounced climate and oceanographic changes has recently been discovered, and like so much else being gleaned from the past, this discovery poses a warning to our world. About 70 million years ago a large plume of hot magma, for reasons yet unknown, detached from deep beneath the earth's surface and began to rise. It reached the earth's surface beneath the Indian subcontinent, which at that time was located off the East African coast in the southern hemisphere. The hot magma spilled out of the earth in one of the greatest and longest volcanic eruptions known in earth history. These great flows are known as the Deccan traps; they cover 6,000 square miles of land in India, and in some areas are over a mile and a half thick. The enormous volumes of magma did not build giant volcanoes; instead they spilled forth in waves over the land surface, emanating from great rifts in the earth. The Deccan traps are one of the largest known flood basalts on the earth; another, younger example, known as the Columbia River basalts, is found in eastern Washington State in North America.

Although the extrusion of these great lava flows certainly affected the local topography (and the local fauna and flora, all of which was burned up by the rapidly advancing flows), the lava itself had little effect on the earth as a whole. But more than lava was brought upward from deep in the earth. Great volumes of volcanic gases, composed mainly of carbon and sulfur dioxide, filled the air. The volcanic vents were like pipelines from hell, spewing fiery brimstone and belching forth the foul, sulfurous breath of the underworld. By the late 1980s geologists had measured the volumes of magma released, which allowed atmospheric scientists to estimate the volumes of gas released into the Late Mesozoic atmosphere; the results of these studies showed that a pronounced greenhouse effect must have occurred. According to various estimates, this volcanically induced greenhouse effect produced a two- to five-degree rise in global temperatures. The changes discovered in Hell Creek plant communities as well as the molluscan extinctions observed in marine sedimentary rocks all began soon after the Deccan eruptions started. Global warming, produced by excess carbon and sulfur dioxide, seemingly led to the start of the Second Event. The most frightening part of this whole scenario is that the calculated volumes of gas vented into the atmosphere by the Deccan volcanic event are quite similar to the carbon and sulfur dioxide emissions being produced by mankind's industries today.

And what of the dinosaurs? Were they victims of these precursor extinctions, as theorized by paleontologist Bill Clemens, or did they hang on until the meteor hit? During the 1980s many paleontologists visited the Hell Creek beds of Montana, trying to better understand the fate of these most prominent of victims. Many different questions were posed. Were the dinosaurs in long-term decline? Some scientists believed that the dinosaurs were already obsolete near the end of the Mesozoic; no catastrophe was needed to snuff them out, since they were on the way out anyway. The major bit of evidence supporting this view came from counts of dinosaur genera found in beds deposited 11 million years before the end of the Mesozoic Era, as compared to the number of genera found in beds deposited a million years prior to its end. Examined solely on the basis of numbers, there was indeed evidence of a profound decline, for the thirty-six genera known worldwide in the older beds had dwindled to at most seventeen genera a million years from the end. This reduction seems unrelated to the climate changes produced by the Deccan volcanics, for it began long before those volcanics began to flow. The apparent dinosaur decline may have been caused by as yet undiscovered factors. However, this supposed decline may have had nothing to do with the actual number of dinosaur taxa roaming around the earth. Paleontologist Dale Russell pointed out that the dinosaur decline may have been more apparent than real, since the data for the higher number of dinosaurs came by combining all known genera collected from over twenty-five localities worldwide, while the younger estimate came from collections made only in one part of one continent, the Hell Creek beds and their equivalents in Montana, Wyoming, and southern Alberta.

Other scientists tried detailed collecting in the Hell Creek strata in order to document whether the dinosaurs underwent a decline during their last million rather than last 10 million years on the earth. Paleontologists Robert Sloan of Minnesota and Leigh Van Valen of Chicago concluded that a major ecological change took place over the last 250,000 years recorded by the Hell Creek strata. They discovered increasing numbers of a new group of mammals, called protungulates, which in their view gradually displaced the herbivorous dinosaurs through competition. Furthermore, these scientists, including Bill Clemens, concluded that the dinosaurs were already totally extinct between 20,000 to 80,000 years prior to the meteor impact. However, a study completed in 1987 came to an entirely different conclusion, suggesting that dinosaurs had survived the KT boundary in Hell Creek. Both of these views were later shown to be false. Detailed study of the very complicated pattern of crosscutting fossil river channels making up the Hell Creek beds showed that the reports of dinosaurs disappearing immediately prior to or soon after the deposition of the iridium-rich clay layers in the region were due to sampling errors.

Two studies completed in the late 1980s and early 1990s finally clarified the patterns of dinosaur extinction. Both relied on enormous data bases. In the first, conducted by David Archibald and Laurie Bryant, students of Bill Clemens, an astonishing 150,000 fossil bones from the Hell Creek region, collected over the past several decades and now curated in the paleontological collections at the University of California at Berkeley, were tabulated in order to measure accurately the severity of the extinction among vertebrate animals. This huge number of fossils was discovered to come from 111 species of vertebrates found in the upper parts of the Hell Creek beds. Of these 111 species, Archibald and Bryant found that only 35 were still alive after the KT boundary crisis: Thus only 32 percent of the vertebrates recognized in this study survived the Second Event. Archibald and Bryant argued that this figure is probably low; some sampling error from the post-Cretaceous part could have biased the survival figure, and survival might have been as high as about 50 percent. But this is splitting hairs. A hideous, rapid, catastrophic extinction suddenly destroyed the plant and animal ecosystems in the ancient Hell Creek landscape. The list of victims and survivors suggests very little pattern of survival, except that larger animals, such as the dinosaurs, suffered most of all. Nineteen species of dinosaurs are known from the upper part of the Hell Creek beds; none survived. Creatures living in the rivers and ponds did better; three of five known crocodile species and six of eight amphibians survived. Contrary to popular opinion, however, mammals did poorly; in the Hell Creek region only one out of twenty-eight species of mammals survived. In all of North America, the survival rate of the mammals was only about 20 percent.

The final nail in the coffin came from a study conducted by Peter Sheehan of the Milwaukee Public Museum. Sheehan trained volunteers to walk the youngest of the Hell Creek beds, hoping they would find large numbers of fossils that could differentiate whether the dinosaurs were disappearing during the last 100,000 years or less of the Cretaceous. Over several field seasons his volunteers recovered nearly 10,000 dinosaur bone fragments. After analyzing all of these fossils, Sheehan found that there was no sign of decline among the dinosaurs during the last years of their reign. Their kind died suddenly, in agony.

7

After a decade of concerted research into its causes, we have a reasonable view of the Second Event. No single cause was entirely responsible for the enormous species death. Volcanic paroxysms spewed huge quantities of gas into the atmosphere, raising the earth's temperature through greenhouse heating and drastically altering atmospheric and oceanic circulation patterns in the process. At about the same time, for reasons still unclear, a great drop in global sea level occurred, further perturbing the climate. And then, about a million years later, an enormous meteor struck the earth. None of these events, even the meteor impact, would probably have wreaked such damage alone; yet through the freakiest of chances, three rare events occurred almost simultaneously. The animals and plants on the earth 65 million years ago went through a terrible patch of bad luck. Half of them paid the price for being unlucky.

Climate change from carbon dioxide emissions, sea level change, and then a meteor. Sixty-five million years later the sequence would be climate and sea level change, the evolution of mankind, and then carbon dioxide emissions. Welcome to the Third Event.

Part Three

The Third Event

Chapter Seven

Autumn

1

The heavy black pistol felt icy and evil in the frigid night air. I aimed as well as possible and gently squeezed the trigger. The gunshot was deafening, echoing across the high mountain valleys. I tried to hand the gun back to the uniformed man beside me; a torrent of Georgian followed by peals of laughter made me turn to my interpreter. Giorgio shrugged. "He says you missed. Try again." Swaying slightly, I lined up the full moon once again in the sights of the military pistol and took another shot, wondering where in the Caucasus Mountains the bullet would finally land. The bright full moon, seemingly unperturbed by my determined efforts to shoot it down with Soviet ordnance, illuminated the craggy mountain peaks surrounding our small party. The Greater Caucasus gleamed whitely in the moonlight but were not yet snow-covered; the ghostly mountains are made of thick white chalks, deposited in a shallow sea soon after the end of the Mesozoic Era. Better than any place on the earth, these strata tell the tale of the planet's recovery following the Second Event.

In the autumn of 1990 I visited the Soviet Union in the company of paleontologist Jan Smit of Holland. We had been invited to view Cretaceous-Tertiary boundary localities in Soviet Georgia, and spent several weeks there looking at a variety of rock types deposited immediately before and after the great, end-Mesozoic extinction. But the trip became more than a simple field excursion to view another KT boundary; in my mind it was to become a metaphor for the aftermath of the Second Event.

We arrived in Moscow on the eve of the Soviet Union's own extinction, in time to witness the climactic death throes of a once-vigorous empire; we then traveled to Georgia, the first of the former colonies to break free of its Soviet chains and the first to realize that following the death of an ancient empire, great, unexpected opportunities for a new and better life are manifest. But the Georgians were also to find that an entirely new regime does not spring fully formed from the ashes of the old; many short-lived kingdoms will rise and fall before a mature, lasting order is once again established. The creaky Soviet empire was finally brought down by the sudden impact of an aborted coup, and the survivors of that great extinction have surely not yet seen the end of the great social changes following such an event. The Georgians are still only in the earliest part of their own Tertiary Period, and like the oceanic and terrestrial survivors of that postdinosaur epoch, they are finding that the extinction of one group of overlords simply opens the door for new groups of tyrants.

2

We met in the Moscow airport in late October, and had to make our way through the city to a different airport for our connection to Georgia. Smit, who along with the Alvarezes should be credited as a cofounder of the impact theory, flew in from Amsterdam; I arrived from London. In the company of a young Russian geologist who served as our guide, we made our way first by bus, then by subway, and finally on foot through the middle of Moscow in an October snowstorm. I felt an unbelievable exhilaration; knowing that an appreciable fraction of the U.S. nuclear arsenal was pointed my way, finally visiting the country I had lived in fear of for so long, seeing the mink caps and endless uniformed men, the towering minarets of the Kremlin and the hulking fortress serving as the KGB building, brightly lighted even late at night, I felt a joy of travel unknown for years. But as my trip progressed and I began to see the incredible dilapidation, the lack of the most elementary technology, the sorry state of transportation and telephones, the lack of food, the squalid cesspools passing as public restrooms, and most of all the dispirited people, I realized the great lie my country had promulgated for so long. My exhilaration gave way to deep anger against those in my own country who for so long taught us to fear the Soviet Union. This country could not feed itself or adequately clothe its people; its cars were antiquated polluters of abysmal design; there were not even cash registers in many stores, and most sales clerks used ancient abacuses. How could the Soviet Union have successfully waged high-tech war against the United States and Europe? As I saw more of the Russian landscape during the ensuing weeks, I began to see the endless scars in the land, the chemical wastes, the unmitigated pollution and garbage far worse than anything I have seen in North America, and I realized that the Soviets had indeed been conducting war, but ugly, low-tech war, directed mainly against their own countryside.

We arrived in the suburban Moscow airport late in the evening, but the snow had increased through the day. Our flight to Georgia had been canceled, and amid hundreds of stranded travelers we were shown some concrete floor space where we could spend the night. Luckily our young guide invited us back to his Moscow apartment, where he and his mother, a chemist at the university, shared their two small rooms. Jan Smit is nearly six foot five and I am well over six feet; the four of us filled the tiny apartment to bursting. Our hosts, both respected scientists, lived on a combined salary of about $200 a month. During our visit there was little food to be had and no wine, but they fed us well on wild mushrooms they had picked in the nearby forests and homemade soup made from potatoes and beets. We talked of science and geology, but at one time during our late-night dinner I made a comment about poor Gorbachev, at that time mired in deepening troubles. One of my hosts vehemently retorted, "Poor Gorbachev, nothing. Poor us!"

3

We eventually flew across Russia and finally crossed the Greater Caucasus Mountains, circled over the Black Sea, and landed in Georgia. The giant mountains to the north and south of this country had long isolated it from its neighbors, allowing the rise of a unique language and culture. Georgia is an ancient country, long ago known as Colchis, the destination of Jason and his fellow Greek Argonauts in their quest for the Golden Fleece. Sitting at a crossroads between Europe and Asia, the Georgians have had long experience with invaders. Buffeted by Christians and Arabs, Mongols and Cossacks, Georgia had most recently been forcefully annexed by the Soviet Union soon after World War I. Georgians counted Joseph Stalin as a native son, the same Stalin who helped turn his birthplace into a slave satellite and abattoir for tens of thousands of his countrymen.

Jan Smit and I arrived on the eve of a historic moment for the Georgians; during our stay they held the first free elections in eight decades, and overwhelmingly rejected their forced union with the Russians. But independence had its steep price. The reeling Soviet state, at that time decaying but not yet quite dead, had staged a bloody incident in the Georgian capital city of Tbilisi several months before our arrival. It had also waged an economic war against the Georgians, with its most effective measure being a blockade of oil imports into the rebellious state. Gasoline was tightly rationed, and of abysmal quality; somehow the roads were still choked with cars, but cars exhaling great quantities of noxious fumes, covering and filling the capital with a sepulchral pall of black soot and smoke.

The Georgians seized independence but were unprepared for it. Their economy was not self-sufficient; they required manufactured goods, oil, and many staples from the evil empire they so hated. And when that empire finally fell into extinction and these goods were shut off, chaos reigned. The rules had suddenly been changed. The shops were empty of food, but full employment required that all take their places tending the empty shelves and racks. We visited a city where an empire had clearly fallen, but the survivors were at a loss about what to do next. The only people flourishing were the human equivalents of weeds, the petty criminals and black marketers who seized every opportunity for profit after the fall of the old order and before the rise of the new.

Our host, the Geological Institute in Tbilisi, had scrounged or stolen enough gasoline to send Smit and me on a cross-country trip to visit rocks of latest Cretaceous and earliest Tertiary age. Along with a handful of Georgian geologists from the institute, we were loaded into an ancient van and set off on our long journey. The crew in our company was a mixture of the old and new. Our driver, Ali, looked like a Mongol, while our chief, a creased geologist named Nodor, was of the old school. Three young, English-speaking geologists were also along, brimming equally with ardent nationalism as well as a thirst for the latest geological discoveries, for news and progress from the western scientific establishments made their way into Georgia painfully, and much delayed. And since we were strangers, we were assigned an official translator as well, a woman who turned out to be as much political officer and guard as translator.

From Tbilisi we traveled far to the west and then followed worsening roads, first into the foothills and then into the heart of the Caucasus Mountains. Tea and grapes are the two major Georgian crops, and both types of agriculture were commonly seen in the lowlands. The broadleaf forests covering the hillsides were putting on a magnificent autumnal display; we had left winter behind in Moscow and were now in a land cloaked in glorious yellow, orange, and red foliage. But as we rose into the higher mountains, the land became much more stark. For six days we journeyed through the mountains, looking at outcrops and studying the rocks. Smit and I saw great and unexpected wonders. We did find new Cretaceous-Tertiary boundary sections, and from one such clay layer, tightly embraced in a wall of white limestone, we found a beautiful graded bed of impact spherules, at that time the best discovered anywhere in the world. (Little were we to know, however, that this wonderful find would be soon overshadowed by Alan Hildebrand's discovery of giant tektites in Central America and Haiti.) We observed the pattern of extinction among both Cretaceous microfossils and larger fossils alike, and with our Georgian colleagues were able to make new studies about the composition of marine creatures that lived in this largely unknown region of the world some 65 million years ago. But exciting as the Mesozoic rocks and fossils were, they paled in comparison to the nature of the Tertiary fossils and strata.

The great meteoric impact ending the Mesozoic Era left behind a very empty world. The oceans were particularly devastated. Great volumes of acid rain made conditions intolerable for creatures with calcareous shells; virtually all of the plankton died off, as did much of the bottom-dwelling fauna. High in the Caucasus Mountains we found great thicknesses of white limestones recording this terrible interval in the history of life. Not only ammonites and microfossils, but large numbers of clams and snails, sea urchins and fish, the commonplace dwellers of the seas, were suddenly extinguished at the base of the Cretaceous-Tertiary boundary clay layers. But the great death was not all that was so well recorded here; the aftermath and the repopulation of the seas was frozen in rock as well.

This great pulse of diversification following the Second Event is beautifully arrayed in the Georgian rocks. New species of bottom-dwelling creatures and swimmers, floaters and recliners, gentle herbivores and savage killers all made their shy or bold entrances into the fossil record as we ascended upward in the Cenozoic Era strata. In boulder-strewn river bottoms or along craggy mountain roads, Smit and I would walk upward from the KT boundaries and watch the repopulation of the seas occur.

The rediversification of oceanic life was a relatively rapid process, but the reconstruction of the ravaged ecosystems was far more time-consuming and proceeded in fits and false starts. Although about as many oceanic species lived as died in the Second Event, the communities in which they lived were shattered. The plankton, the benthos, the pelagic ecosystems: All were ravaged or destroyed as species after species comprising the interlocking energy links became extinct. Even if many creatures survived the catastrophe, their communities did not, for the flow of energy through these ecosystems was surely disrupted by the disappearance of so many key links. It would be like asking a building to stand when half of its bricks are somehow suddenly and randomly removed; the rest certainly comes crashing down as well.

Climbing upward through time, we could see the first appearances of the newly evolved. Some, such as new species evolving from a lone surviving nautiloid cephalopod, appeared with a flourish in the oldest Cenozoic rocks. Other groups, such as the corals, reappeared far more slowly; following the meteor impact, at least 10 million years would pass before coral reefs once more graced the earth. The earliest communities of the Cenozoic Age preserved in these Georgian strata look little like today's, for many false starts and failed experiments mark the recovery period following the Second Event. But as our party climbed ever higher into the Caucasus Mountains, passing simultaneously into ever-younger rocks, we could watch the fossil assemblages become increasingly familiar as they became dominated by the creatures characteristic of our time, such as modern clams and snails, sand dollars and diatoms, bony fish and crabs. A new world, our world, eventually evolved in the wake and waste left behind by the dinosaur-killing asteroid.

The youngest rocks Jan Smit and I ultimately observed in Georgia were about 35 million years old; they were deposited in lakes rather than deep in the sea, and because of this the fossil content had changed markedly. We found the remains of fish scales and freshwater mollusks, but by far the most common fossils were from plants. Most of the fossils were the stems or leaves from trees, but others were less conspicuous; here and there a small grass blade was preserved, or the branch from a low-growing herb. Looking at these fossils in the dim light of a gray, late-autumn day, high in the Caucasus Mountains, they reminded me that the meek really had inherited the earth.

4

Seattle, my hometown, has always been a great moviegoer's city; numerous small cinemas defy economics and compete with the giant multiplexes—and get away with it. One of the art house favorites here (and one of my personal favorites as well) is a beautiful gem of a movie called *King of Hearts*. Alan Bates stars as a reluctant soldier ordered to enter an empty, small French town that has been filled with explosives by retreating German troops. In the face of imminent destruction, all the townspeople except those locked up in the insane asylum have fled. The gates of the asylum are left open by the retreating townspeople, and the long-incarcerated lunatics—who, of course, turn out to be the only sane people in this beautifully poignant antiwar film—gradually escape, to wander free for the first time in the great, empty town they had long but so little inhabited. But the newly liberated residents of the asylum don't immediately realize they are free, for some time goes by between the actual act of their liberation—the disappearance of the townspeople and the opening of the asylum's gates—and the inmates' discovery of this fact. And second, even with the gates open, the lunatics take some time before they *decide* to venture out into the strange, new, empty world. But when finally they do, a revolution rapidly occurs. In the asylum, all of the residents wore the same clothes and acted in the same way. But once liberated into the empty town, they each head off in a different direction and soon take up the manners, clothes, and professions of the town's former inhabitants, as well as commencing entirely new jobs. One man dons the robes and offices of a priest, another becomes the town barber; one woman reopens the town bordello, while another assumes the finery and manners of the aristocracy. Soldiers, firemen, members of an orchestra, a rugby team; each of these professions is soon resumed—but not exactly in the same way as by the village's previous inhabitants. The new barber pays his clients to get a haircut, the firemen set fires rather than put them out—the interpretation of old jobs by a new and vastly different set of characters subtly (or not so subtly) changes how old occupations are administered.

The *King of Hearts* parable reminded me of what surely happened after the death of the dinosaurs. During the long Mesozoic summer, the dinosaurs were undisputed monarchs of the land. Mammals existed as well, but were all small in size and quite inconsequential components of the fauna. Most of the Mesozoic mammals were mouse or rat size and probably existed on a diet of insects or vegetation. Many possessed extremely large eye orbits in their skulls, adaptations suggesting that they possessed the large eyes of nocturnal creatures, thus allowing them to avoid the brunt of dinosaur predation. The Mesozoic mammals were essentially incarcerated by the dominance of the dinosaurs, imprisoned in the tight corners and poor fringes of the ecosystems as surely and as securely as any assemblage of lunatics locked within an asylum; they were there, but played little or no part in the course of affairs. And then, suddenly, the dinosaurs disappeared from the world. I wonder how long it took before the surviving mammals, now among the largest creatures left alive on land, finally realized that they were the new masters? For how long did the nocturnal habits of the tiny mammals continue, even when no longer necessary, simply because they had been so long living in that fashion? How deeply ingrained in our furry ancestors' DNA was the need for furtive secrecy, to live in tiny burrows, or high in trees, to feed on the world's refuse rather than more delicious delicacies, simply because those delicacies were found only out in the open, where swift death had so long ruled? When did the trembling mammals finally creep out of their nocturnal asylum to find that the old villagers and rulers were gone?

When the mammals finally did leave their burrows and dark corners, they found a rich, empty world to exploit and many new professions to assume. Vegetation had come back quickly, so the world was still redolent with luxuriant jungles, and these, along with the many surviving insects, frogs, lizards, and even their own kind, presented great opportunities for mammalian herbivores and carnivores, swimmers and fliers, larger grazers and long-necked browsers—the many feeding methods once employed by the myriad dinosaurs. Over a 5- to 10-million-year period, bursts of evolution allowed the early Cenozoic mammals to put on these new costumes, as new teeth and body shapes evolved to allow exploitation of the earth's suddenly available treasures. Once the old villagers were gone and the locked gates finally and irrevocably opened, some 65 million years ago, the former inmates became kings.

5

The Cenozoic Era is often called the Age of Mammals (although it should perhaps be characterized as the Age of Beetles, since these bugs are the single most speciose group of organisms on the earth). But since those doing the describing (us) are mammals as well, the Age of Mammals it is.

The first mammals evolved from protomammals about 200 million years ago. The key to mammalian success is usually attributed to our warm-bloodedness, parental care (including the suckling of young), and an upright rather than sprawled gait. All of these traits are no doubt important factors in the undoubted success of our class. But the single most important factor may be the nature of our teeth. Almost all mammals have teeth that occlude, or meet face to face, thus allowing them to chew their food. Mammalian teeth come in a bewildering array of morphology, reflecting the wide spectrum of food eaten. Mammal teeth are so distinctive that paleontologists usually can identify the correct genus of fossil mammal from a single molar or premolar. As the mammals evolved, their teeth changed rapidly to take advantage of new food opportunities.

Yet, however wonderful mammalian design was and is, there is just no getting around the fact that for two-thirds of mammals' existence on the earth, they were small in size and few in number and diversity, so thoroughly were our Mesozoic ancestors dominated by the dinosaurs. It is probably a safe bet to assume that if the dinosaurs had not had the bad luck to be wiped out completely by a random asteroid from space, we would still be rat-size creatures hiding in trees. During the Jurassic and Cretaceous periods, the heyday of the dinosaurs, there were between ten and twelve lineages of mammals, all about the size of rats or smaller. Many showed adaptations for living in trees, and many had pouches, suggesting that they had reproductive strategies similar to those seen today in marsupials.

By the end of the Cretaceous Period, there were three major types of mammals: monotremes, which lay eggs (the duck-billed platypus is one of the last living representatives), marsupials, and placentals, which represent the vast preponderance of currently living mammals, including mankind. The great success of the placentals, however, is a post-Mesozoic phenomenon; the marsupials were by far the most successful group of mammals during the Late Cretaceous.

The Second Event was harsh to warm-blooded creatures. If we agree with the dinosaur specialists' assertion that dinosaurs were warm-blooded, warm-blooded animals suffered much worse than cold-blooded creatures during the mass extinction at the end of the Mesozoic. Most cold-blooded reptilian lineages, such as the lizards, snakes, turtles, and crocodiles, survived with only moderate losses. But all of the dinosaurs disappeared, as did most of the mammals. (The fossil record of 65-million-year-old birds is virtually nonexistent, so we have no evidence about how these warm-blooded creatures fared in the Second Event.) The mammals and the dinosaurs were greatly affected; in the Hell Creek region of Montana, for instance, only one out of thirteen species of marsupials survived the event.

The best record of the earliest Cenozoic mammals comes from the same place as the best, last record of the dinosaurs: the Hell Creek region. The oldest Tertiary beds there are the Fort Union Group, and it is in these variegated, coal-laced strata that the mammalian rebound following the Second Event is best displayed. Between 65 and 58 million years ago, the surviving mammals diversified from only three families to over forty. Within about 12 million years following the extinction, the few survivors had evolved into well over 200 genera and perhaps thousands of species. By 50 million years ago, most mammalian families now present on the earth had evolved, and many short-lived mammalian empires rose and fell as the terrestrial ecosystems attempted to reach some new stable equilibrium.

Paleontologist Michael Benton of England has written that during the first few million years following the Second Event, the fauna of the earth might have looked much like the animals found today in urbanized parts of America or Europe, minus the human beings and our domesticated animals, such as dogs and cats. In any city you can see many small birds of a few different species, an occasional lizard or snake, and if you prowl the parks, garbage heaps, or the assemblage of road-kill, a few small mammals. In my city, Seattle, an occasional opossum or raccoon can be seen amid the mammalian fauna of shrews, rats, moles, and mice. In size, all of these modern-day city dwellers resemble the Early Cenozoic assemblage of mammals. Only the opossums and shrews, however, look anything like actual mammalian species existing over 60 million years ago.

Following the extinction, 65 million years ago, many of the surviving mammal species rapidly increased in size or evolved new species that did. By about 55 million years ago, archaic, rhinoceroslike herbivores roamed the land, pursued by primitive, dog- and bear-size predators. These first experiments in larger herbivores and predators would seem extraordinary to us, if we could see them alive today, for they looked nothing like the current crop. The largest of these giant herbivores has been named *Uintatherium;* it was like a gargantuan rhinoceros and had six bizarre horns on its misshapen head. Coexisting with and chasing after the herds of uintatheres were strange carnivorous species. One of the earliest groups of carnivores evolved from hoofed herbivores; there is no modern analog to these giant, hoofed predators. (Imagine a large carnivorous horse chasing after you, snorting and smacking long pointed teeth. Sugar cubes, apples, and saddles would be inappropriate.)

Living among these bizarre forms were many smaller species as well, for many evolutionary recipes were cooking. The first primates were climbing into the trees, while the first rodents, bats, and the earliest ancestors of most modern mammals can be found among these Early Cenozoic assemblages within the Fort Union beds.

The turning point in mammalian evolution, when the modern-day fauna finally displaced the first, archaic assemblages, coincided with the cooling of the earth 40 million years ago and the spread of grasslands. At this time the herbivorous ungulates, now our world's dominant mammalian herbivores, split into two groups, the odd-toed forms (which include horses, tapirs, and rhinos) and the even-toed group (comprised of pigs, hippos, cattle, deer, giraffes, camels, and antelope).

Modern carnivores appeared as well, diverging into the feliforms (cats, hyenas, and mongooses) and caniforms (dogs, bears, raccoons, weasels, and seals). By this time primitive whales and other marine mammals had already invaded the seas, and bats began competing with the birds for mastery of the skies. But all of these new mammals were originally adapted to a world much like the one known to the dinosaurs, a warm world dotted with steamy swamps and humid tropical jungles. By about 45 million years ago, the warm, continent-covering jungles began to recede from many lands as the earth cooled. The mammals' first great burst of evolution, following the death of the dinosaurs, was in response to a suddenly empty world. The second great burst, some 40 million years ago, occurred when grasslands and low herbs began to replace the trees. Many of the earliest-evolved groups of mammals began to disappear at this time, to be replaced gradually by forms better equipped for life on the open grasslands. Thus the modern-day deer and bovids appeared, as did giraffes, elephants, pigs, and horses. New, swifter carnivores with greater intelligence were required to catch these larger, fleeter herbivores of the grasslands. The formation of grasses also favored the evolution of small creatures as well as the large. Rodents, with their teeth exquisitely adapted for eating a variety of seeds and grains, showed their greatest proliferation following the rise of the grasslands and herbs; in turn, the diversification of these small mammals initiated a great diversification among the snakes. Large numbers of passerine, or perching, birds came into existence at this time as well. Increasingly, the creatures of our world took on a modern appearance, and all the while the earth continued to cool.

6

By mid-November the freezing wind had left crusts of ice on the pools and ponds, and the last birch leaves had fallen away from the treed riverbottoms in the Caucasus of Soviet Georgia, a country on the verge of becoming the Republic of Georgia. Jan Smit and I had entered territory long forbidden to foreigners, but the crumbling Soviet empire could no longer exert its once-absolute, iron control on all tourists in the USSR, and our Georgian hosts had defied protocol to take us into particularly rich outcrops near a small town north of the regional capital of

Kutaisi. We stayed there for three days, looking at outcrops by day and freezing by night. Our unheated hotel had only one room suitable for "tourists," and it had but a single bed and no hot water. Despite all of our protests, Jan Smit and I found ourselves sleeping in this bed together, with both of us wrapping our feet in layers of socks each night, since they stuck far out from the end of the bed. Our presence surely attracted attention in the small, close-knit town, and on our third night a loud pounding on our hotel door announced insistent guests. Two dark men entered and brusquely demanded our passports, which we just as insistently refused to hand over. A rather heated argument ensued, with neither side speaking more than a few words of the other's language. Our translator luckily appeared, and another heated argument broke out, with Nodor, the chief Georgian geologist, finally being summoned. The two men finally left, but glowered threateningly at us as they did so. Afterward, our Georgian colleagues would not discuss the incident, nor tell us who the men represented. We were to later find that Georgia is not a single, happy country but is made up of many tiny regions, each with its own politics and all distrustful of those from the capital of Tbilisi. After a strained dinner, with our hosts drinking even more than their usual bottle of wine per person, Smit and I resolved that it was time to get some air and get away from the claustrophobic embrace of our Georgian baby-sitters and their arcane politics.

Following dinner we surreptitiously strolled out of the hotel and walked into the center of town, finally finding a small coffeehouse still open. We were sitting there, nursing weak tea, when one of the young, English-speaking Georgian geologists came bursting in. He looked immensely relieved to see us, and began haranguing us for disappearing. "Relax, Giorgio," Smit admonished, but Giorgio was not to be placated. We ordered him a tea and finally calmed him with stories about the wide world to be found outside of Georgia's borders.

Thus engaged in a pleasant evening, for once out of our confining hotel room, our small party of three was unexpectedly joined by new guests. I saw a shadow fall across Giorgio's face, and looked up to see three large men standing beside our table. A barrage of Georgian was directed at us. I asked Giorgio who our visitors were, and he replied that it was the chief of police, his deputy, and some functionary known as the town "procurer." (This latter title elicited raised eyebrows on my part.) After some discourse the chief looked at us and smiled. "He wants to buy you a drink," Giorgio told us. "He has never seen a real live American before." I replied that I would be delighted. After translation, Giorgio told us that we would have to follow the police to get our drink. We were hustled out of the coffeehouse and into the backseat of a large black sedan of indeterminate ancestry. The police chief took the wheel, and we sped off into the dark night, heading into the mountains. Increasingly nervous, I asked Giorgio why we couldn't just stay where we were, drink a toast, and head off on our separate ways. Giorgio replied that the chief had decided that there would be no further drinking in his town that night, since the following day was to be the official election day, and he didn't want anyone too hung over to vote. With that, we headed for the nearest village where liquor could be procured, one steep mountain pass away.

And so we ended up high in the mountains, drinking wine, and ultimately shooting pistols in honor of new friends and falling empires. Upon our return (which surely was the most dangerous trip of my life, for the police chief was negotiating hairpin turns while being absolutely *lit* from the Georgian wine), Smit and I were astonished to see the geologist Nodor, followed by our official translator and the rest of our Georgian companions not safely in bed at this late hour but walking the streets of town, searching for us. Our translator was particularly distraught, thinking, no doubt, of her fate had she managed to lose the two foreigners in her charge. My last view of this memorable night was seeing Giorgio pushed into his room by an apoplectic Nodor, undoubtedly to receive the tongue-lashing of his life.

The following day the long-awaited vote took place. The great Georgian churches served as the balloting stations, and the people turned out in droves to finalize the plebiscite confirming their will for independence. The Nationalist Party won by a landslide, and the Communist Party was routed. A stout Georgian named Zviad Gamsakhurdia was elected president, and a new dawn was proclaimed. But like the first ecosystems arising after the great extinction ending the Mesozoic Era, this presidency was but a stopgap, and all too temporary. Within a year the same guns recently aimed at the moon would help force the first elected president of the Georgian Republic into exile.

Chapter Eight

Winter

I

The world cooled and dried; like falling leaves, the great tropical forests fell away from the land. Our ancient primate ancestors found themselves rocked out of their forested cradles and came tumbling down out of the trees to face their test of winter.

By 6 million years ago, the world had taken on a nearly modern appearance in many respects. The drifting continents had reached the approximate positions they occupy today; most of the world's great mountain ranges were in place. But several seemingly innocuous changes were under way, changes that would drastically alter the world and ultimately decimate its biota. One of these changes resulted in a worldwide drop in sea level and the complete desiccation of the once and future Mediterranean Sea. That event eventually helped force our primate ancestors out of the trees and make them humans—and hunters—in the process.

The great climate changes that have affected our earth over the last 6 million years are themselves largely the products of continental drift. The continent of Antarctica, one of the old, founding members of the now-ancient Gondwana supercontinent, had remained associated with its southern cronies South America and Australia, even after other members of the huge landmass had pulled away. But during the mid-Cenozoic Era, about 30 million years ago, this long marriage ended as Australia began to drift northward and South America continued a westward drift. Antarctica, left on its own, began slowly to drift southward, until it was centered on the South Pole. The isolation and southward drifting of Antarctica made the continent unprecedentedly cold, for no warm ocean currents could now bathe any of its shores. Each year winter on the Antarctic continent became longer and colder. Snow accumulated during the winter months never completely thawed by the end of the following summer, and in this fashion a great ice cap began to grow. Prior to this time, there had been no polar ice caps, for continental landmasses had not been sitting over either of the earth's poles for 250 million years, and although pack ice can cover cold oceans, true *ice caps,* which are great accumulations of ice hundreds of feet thick, cannot form on open sea.

The growth of an ice cap on the Antarctic continent 6 million years ago drastically influenced the earth's weather. As more and more bright-white ice accumulated on Antarctica, it caused increasing amounts of sunlight to be reflected back into space. A vicious circle began; as the huge, white ice cap grew, less sunlight warmed the earth, and the planet then became even colder, making the ice cap grow bigger. Made of water that originated in the sea, as it grew, the level of the world's oceans fell. Like water draining out of a bathtub, the level of the seas everywhere on the earth dropped by 150 feet.

Far to the north of Antarctica, the Mediterranean Sea became cut off from the Atlantic Ocean by the drop in seawater. The Mediterranean was a vestige of the Mesozoic, world-spanning ocean known as Tethys; with the formation of the Alps, however, and the collision of Africa with southern Europe, the Tethys Sea contracted in size, until its connection with the open Atlantic was through a single, shallow passage now known as the Straits of Gibraltar. In this fashion the Mediterranean Sea was born. With the rapid drop of sea level, the Mediterranean became completely landlocked. In bright, hot sunlight, it began to evaporate.

The discovery that the Mediterranean Sea completely dried between 6 and 5 million years ago was made only two decades ago. The then newly instituted Deep Sea Drilling Program discovered that the entire sea floor of the current Mediterranean Sea is underlain by a giant salt deposit, which could have formed only if the entire volume of seawater held within the sea's gigantic basin evaporated entirely and relatively quickly. The results were certainly catastrophic for its inhabitants. All of the myriad sea creatures in the Mediterranean at the time must have met a hideous death, for as the large sea slowly shrank, it became increasingly saline, until it was but a large, briny bath, not unlike Utah's Great Salt Lake. Eventually the entire former sea was turned into a great salty desert.

The effect of the Mediterranean's desiccation on the surrounding territory must have been profound. Oceans always exert an ameliorating effect on local climate, causing nearby land areas to receive rainfall and buffering great temperature swings. When the Mediterranean disappeared, a great region of the earth's surface that had previously received the benefits of a maritime climate suddenly changed, becoming hotter and drier. The immediate impact was a rapid change from wet forests to dry grasslands. The myriad creatures living in the rapidly disappearing forests had three choices: migrate, adapt, or die. In the previously forested regions of east Africa, one group of primates found itself suddenly thrown out of the life-giving, life-sustaining trees. No longer could these primates hide from great predators or enjoy the arboreal largesse that had sustained their kind for more than 50 million years. On the newly formed African grasslands, they needed greater size and better locomotion but, most of all, an ability to outwit their numerous predators.

The drying-up of the Mediterranean helped create the beginning of a new age, the Age of Humans. We were driven from the Garden of Eden by desiccation, dry winds, diminishing rainfall, and the changeover of our womblike forests to the dry grasslands of Africa.

2

The earliest primates, our first ancestors, may have existed prior to the end of the Age of Dinosaurs, and certainly did so soon after the dinosaurs' extinction. These creatures looked much like shrews, and probably acted in rather the same way; they lived in trees, and their dentition suggests that they ate insects. During the early part of the Cenozoic Era, between 65 and 50 million years ago, many early primate species existed around the world, all showing a variety of characteristics distinguishing them from other mammals. Some of these features, such as grasping hands, feet, and mobile shoulder joints, are clearly adaptations for living in trees. The nature of the head region, with its flat face, acute and forward-facing eyes (thus yielding binocular vision), and a relatively large brain, are also features that may have evolved in response to an arboreal life-style. Natural selection acts quickly when your life depends on first seeing and then catching branches as you brachiate through the trees, high above the forest floor, especially when one mistake can be fatal. Primates show increased parental care compared to most other mammals, and a long period for raising the young; a further consequence of this is a very low birthrate, involving but one or two young per pregnancy. This too may be related to the dangerous life of living high in the trees, for the young must be watched carefully until they can master the hazards of a high-wire, netless life-style.

For the first 10 million years of primate history, most of our ancient ancestors looked much like the modern-day tarsiers or lemurs. About 40 million years ago, however, a new group arose: the monkeys. As the world cooled and forests increasingly gave way to the grasslands now so typical of our world, the primates either had to adapt or disappear. The primates disappeared from North America, a region where the Early Cenozoic primates had been particularly common. In the latter parts of the Cenozoic Era, primates became largely restricted to tropical forest environments in equatorial regions.

The first of the apes had evolved by about 20 million years ago. Paradoxically, although this group is the most intensively studied of any mammalian taxon, the nature and evolutionary interrelationships of the Hominoidea (Apes and Humans) is still highly problematical. This confusion arises in no small way from the very incomplete and spotty fossil record of apes and humans. Our skeletons, and those of our ancestors, didn't routinely enter the fossil record. In spite of having bones, we rarely fossilize.

The earliest of the apes is named *Proconsul* (which is a great, whimsical name; apparently there was a famous chimp named Consul, living in the Manchester Zoo in the 1930s. When the fossil bones of the earliest ape were described in 1933, the species they came from was named *Proconsul*, or "before Consul," an understatement if there ever was one). *Proconsul* was about the size of a baboon, ate fruit, and lived in trees. It certainly walked on four legs if ever it visited the ground, and it could have been the common ancestor of all subsequent apes and humans. A great variety of forms appeared soon after *Proconsul,* and with the continental collision of Africa with Eurasia, about 18 million years ago, these creatures soon spread from their African birthplace to Asia. From African fossil beds deposited between about 20 and 15 million years ago, we have a rich record of these creatures, detailing a great evolutionary radiation of apes. And then the fossil record, at least for the traditional hominid hunting grounds of eastern Africa's Great Rift Valley, almost disappears. A few fossil hominoids are known from this area in sedimentary rocks aged between 14 and 4 million years ago. The fraternity of hominoid bone hunters is keenly frustrated by this hiatus in our knowledge. Happily, the gap has been bridged at least slightly by a recent discovery in southern Africa.

In March 1992, a consortium of scientists announced in *Nature* magazine that they had found a 13-million-year-old jawbone from Namibia, a small country bordering South Africa. The presence of this fossil in Namibia was a pronounced surprise, for most physical anthropologists believed that early hominoid evolution largely took place in what is now Kenya and Uganda, and not in the southern parts of Africa. The fossil in question, found in the limestone tailings of a mine, was the only hominoid material found by a large team of hunters. But the single bone represents a large section of a lower jaw, with most of its teeth intact. There is enough material to show that it represents a probable common ancestor of apes and humans, and may be the species that ultimately gave rise to these two diverging stocks of primates, a split that took place between 5 and 10 million years ago. More fossil material is needed to place the fossil more accurately in our family tree.

Africa was largely forested as late as about 15 million years ago, but about that time its great tropical forests shrank. Northern Africa gradually became drier, while lands to the east and south became regions of savannah and scattered trees. In this world the primates continued to evolve. Most modern monkey groups appeared in a widespread evolutionary diversification occurring about 8 million years ago. But the long, largely closed 10-million-year period, when we have but the single jawbone from Namibia to show 10 million years of hominoid evolution, is maddeningly frustrating; during that period great changes must have been taking place, changes about which we can only speculate. In 4-million-year-old sediments, fully bipedal hominid fossils have been found. The oldest species is formally known as *Australopithecus afarensis*.

After this time, the rise of humanity was apparently swift. The spectacular discovery of a nearly complete, 2.6-to-3.2-million-year-old hominid from Ethiopia by anthropologist Don Johanson, a fossil of a young female he affectionately named Lucy, filled in many of the missing gaps in our knowledge of human evolution. Lucy and her kind represent the oldest member of our tribe. They were far smaller than us, with the largest males weighing about 100 pounds. One of the odd aspects was the striking sexual dimorphism: Males were between 50 and 100 percent larger than females. This feature suggests that, like many modern primates, the australopithecines traveled in troops like baboons rather than forming permanent family groups. The brain of these creatures was about 20 to 30 percent larger than that of a chimpanzee and about 33 percent of the size of ours. It is apparent from fossil records that a fully bipedal existence preceded a large brain size. The australopithecines may have been a bit like the character in a wonderful cartoon penned by the late artist Kliban: great dancer, but not much good at algebra.

Several species of australopithecines existed between 2 and 3 million years ago in Africa. One of these probably gave rise to the first member of our genus: *Homo*. Primitive members of our own genus are differentiated from the australopithecines by a larger brain capacity, improved bipedal locomotor ability, and a shortened face. The oldest species of *Homo* has long been considered to be *Homo habilis*, discovered by the great paleoanthropologists Louis and Mary Leakey. In 1992, however, I was gratified to see a wire service newspaper article reporting that an even older *Homo* had been recognized, found in beds almost 2.5 million years old, and described in the journal *Nature* by a pair of anthropologists named Hill and Ward. It is wonderful when your brother is a scientist.

3

Why both Ward boys became paleontologists still baffles my family. But from early boyhood onward we were both mad for fossils, and our respective paths paralleled the two main branches of paleontological study. I became a geologist interested in fossils, while my brother became an anatomist interested in vertebrate evolution.

The earliest paleontologists were all anatomists. Great men such as Georges Cuvier, Richard Owen, Thomas Huxley, O. C. Marsh, and Ray Lankester were all students of comparative anatomy and used their detailed understanding and knowledge of modern creatures as a key to the past. Most paleoanthropologists (those who study the fossil record of humanity) and some vertebrate paleontologists (students of the backboned animals) still have backgrounds in anatomy. But today most paleontologists studying the fossil record are trained principally in geology, and the great majority of them end up working for oil companies rather than conducting publishable scientific research. My brother and I ended up on different sides of this schism. While I was studying stratigraphy, my brother was immersed in *Gray's Anatomy;* when I was measuring sedimentary rocks in geological field courses, he was carving up cadavers. In the best of all worlds, each of us studying the history of life and its fossil record would benefit from both pathways, but no life is that long. Those of us arriving in paleontology via geology departments must pick up a knowledge of anatomy on the side, while anatomists must spend their spare time poring over geology texts. Neither avenue is right or wrong, better or worse; both, however, usually leave their protagonists weaker in one area. Yet aside from training, there is another great difference that sets the hominid seekers apart from all other paleontologists: They all seem to come out of graduate school with a press agent in their pocket.

It is no accident or wonder that the human bone hunters sit atop the paleontological pyramid and represent one of the most glamorous branches of science—any science: All of us, to some degree, are curious about our origins, and the average person is going to be far more interested in the evolution of humanity than the evolution of pygidial appendages in Cambrian trilobites, or the ontogeny of phylloid sutural elements in upper Cretaceous ammonites (one of my expertises). But because the fossil record of our ancestors is abysmal, many different interpretations can be made from the usually meager, often equivocal evidence. On a less generous level, the political and financial stakes are high: Millions of dollars of grant money is at stake, both in government research money and, to a greater extent, from private philanthropists. I don't know a single invertebrate paleontologist (other than those who fortuitously found a few oil wells) who ever became rich from digging fossils. On the other hand, I know of several paleoanthropologists who have become very wealthy from their work.

The current armed camps are roughly divided into a Richard Leakey side and a Don Johanson side. It is therefore astonishing that my brother has so gracefully and, up until now, so anonymously glided between the two factions, while nevertheless making some provocative and important discoveries. At family gatherings, he would casually tell stories of exploring the Great Rift Valley in Africa with Richard Leakey, or prospecting for Miocene apes in the Siwalik Mountains of Pakistan. What he wouldn't discuss with us, or with reporters, were his own discoveries. Yet in the early 1980s he convinced everyone in his field that an early ape named *Sivapithecus* was not, as heretofore believed, a member of the human family tree, but rather the base of the branch leading to the orangutans; this discovery required the revision of all textbooks discussing the course of human evolution. This and other solid work was built from patient, unglamorous but absolutely necessary anatomical description and analysis. Although it is far more fun to go into the Great Rift Valley and find a hominid than to sit in a small lab and figure out what it is, the latter endeavor changes fragmentary bones into parts of our family tree. Sadly, the unglamorous anatomical work is quite often overlooked following a new fossil discovery. But in early 1992, analyses of material collected several years ago moved my brother out of the scientific shadows and onto center stage: Steven Ward and Andrew Hill pushed back the age of the first member of our genus, *Homo,* by more than half a million years. Their discovery showed that the first species of *Homo* existed almost 2.5 million years ago; it also vastly complicated the taxonomy of the early hominids, for it showed that there may have been several species of *Homo* running around the African countryside in the critical half-million years between 2.5 and 2 million years ago. The 2.5-million-year-old date also may clear up another mystery. The oldest known stone tools date from rocks of this age; they had long been attributed to the australopithecines. But to many anthropologists, it now seems likely that these most ancient of artifacts came not from the small-brained australopithecines but from the newly evolved *Homos*, for the australopithecine hominids were present on the earth as early as 4 million years ago, yet stone tools first appear about 2.5 million years ago. Perhaps tool-making, as much as any anatomical feature, will someday be recognized as a unique characteristic of *Homo*.

Two and a half million years ago. It is a date of significance to many different branches of science. To paleoclimatologists it signifies the onset of the great climatic perturbation that began the Ice Ages. To anthropologists it now marks the first appearance of our genus and a great diversification of hominids. To geologists it denotes the end of one geological time unit, the Pliocene Epoch, and the start of another, the Pleistocene. I believe that it also marks the start of the modern extinction, the Third Event.

4

The great drop in sea level occurring 6 million years ago, the event that caused the Mediterranean to dry up, lasted only about a million years. This period (known as the Messinian Event) ended when the Antarctic ice cap temporarily receded and its melting ice caused the oceans to rise in their basins. For about 2 million years thereafter the earth enjoyed a spell of warmth, the last days of our long autumn. Southern England sported a subtropical flora, and Iceland was a relatively warm, pleasant place. It was not to last. When the body of the *Homo habilis* individual ultimately to be described by my brother was falling into its gritty African grave some 2.5 million years ago, a change far more precipitous and sweeping than any before was rapidly overtaking the earth's climate. It was the start of the Pleistocene Epoch, the Great Ice Age. Winter had arrived.

The realization that the earth was profoundly affected by a long period of cooling, resulting in the periodic growth of numerous gigantic ice sheets that covered vast regions of the northern hemispheres relatively recently, was a triumph of nineteenth-century science. It was a discovery that did not come easily or unopposed, however. European naturalists had long recognized that the many large boulders dotting the

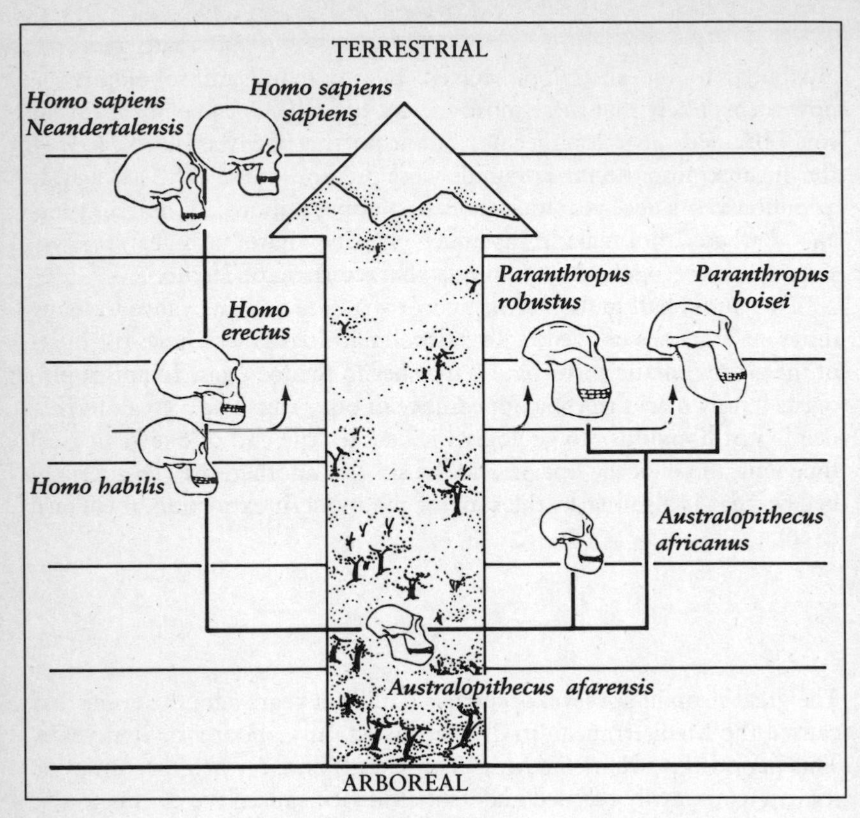

Evolutionary tree of humanity, during the Pliocene and Pleistocene epochs.

landscape, even in areas far from their probable origins, were extraordinary occurrences; when gigantic boulders of granite are found in regions hundreds of miles from the nearest source of such rock, some explanation is required. The easiest excuse was that the Flood did it; during the great Noachian deluge, according to biblical reasoning, the rushing, world-covering waters carried many large boulders to their present, perched positions. It took the Herculean efforts of a great geologist named Louis Agassiz finally to demonstrate that these boulders, as well as many other curious features, such as U-shaped valleys, gouges and scratches on rocks, and great piles of gravel scattered across the northern parts of Europe and North America, had been caused by the movement of continental glaciers.

Starting in the polar regions and then slowly, inexorably spreading outward toward the warmer regions of the earth, these great sheets of ice completely changed the nature of life on the planet and, in many regions, the geography of the earth itself. In North America alone, the Great Lakes in the midcontinent, Puget Sound in Washington State, and the great inside passage stretching from southern British Columbia to Alaska were carved out of solid rock, while a huge pile of gravel and debris was spread over large expanses of the continent. The great ice sheets were over a mile thick in most places, and even the regions not covered by ice were in some way affected, for the earth's climate was radically changed and turned topsy-turvy. Gigantic, cold deserts and semideserts expanded in front of the advancing ice sheets, while regions normally dry, such as the Sahara of northern Africa, experienced increased rainfall. Conversely, the great rain forests covering the Amazon Basin and equatorial Africa, regions of relative climatic stability for tens of millions of years, experienced a pronounced cooling and drying and became dotted with savannahs and open regions due to the rapid, worldwide climate change brought on by the glaciers.

It had long been postulated that the Ice Ages were composed of four separate glacial advances and retreats during the last 2.5 million years, with the last ending only a little more than 10,000 years ago. Recently, however, improved chronology (based on oxygen isotopic ratios, derived from planktonic microfossils) coupled with many new radiocarbon dates from terrestrial sites have shown that the history of glaciation was far more complex. At least eighteen separate ice advances and retreats are now known, occurring roughly at 100,000-year intervals, with the severity and size of the glaciers produced during each cycle increasing through time.

As might be expected, the last of the glacial intervals left the best geologic record. This last glacial advance and retreat, known as the Wisconsin glaciation in North America and the Würm in Europe, began about 35,000 years ago and ended about 10,000 years ago. At its maximum extent, the Wisconsin ice sheets covered most of Canada and extended far to the south in the American Midwest, while great glaciers also began and grew out of the world's high mountains. England, Scandinavia, Greenland, and much of the Baltic region of northern Europe were also buried under a mile of ice.

The cause of these great glacial advances and retreats has long been debated. The primary agent is easy to pinpoint: The earth became colder. As during the earlier Messinian Event of 6 million years ago, the great ice caps began to spread from the polar regions when the earth cooled. As the ice caps spread, the amount of sunlight reaching the earth was increasingly reduced. But why was there a series of pronounced cycles, and why did the ice caps start to grow in the first place?

Several explanations have been favored. Some scientists believe that the sun's energy output diminished, while others point out that the closing of the Isthmus of Panama, which occurred at about the same time, radically changed oceanic circulation patterns and brought about a period of cooling in the process. Most scientists, however, suspect that the glacial advances and retreats have a more astronomical cause, with the changing distance between the earth and sun the culprit.

The earth travels in an elliptical rather than a circular orbit around the sun. But the spinning earth is also like a giant gyroscope, and, like that toy, it slowly wobbles as it spins. The tilt of the earth's axis (which causes the seasons) is what wobbles as we revolve around the sun and also slightly varies over long periods of time, bobbing up and down between about 22 degrees and about 24.5 degrees over a cycle of about 41,000 years; the axis itself slowly revolves, with one revolution taking 22,000 years. As a result, the severity of summers and winters will change gradually, depending on the relationship between the earth's tilt and its distance from the sun. Summer in the northern hemisphere is likely to be hottest when the longest day of the year coincides with the earth being at the point in its orbit closest to the sun. The earth is in this position every 22,000 years.

This changing relationship is called precession of the equinoxes. According to a theory first proposed by a Yugoslav astronomer Milutin Milankovitch, the glacial advances were set off when the winters were coldest during the 22,000-year precession cycle, when the earth was at its maximal distance from the sun. But there must be more to the story, for precession has been occurring for as long as the earth has orbited the sun, while ice ages have occurred infrequently: 400 million years ago, about 275 million years ago, and starting 2.5 million years ago. The drift of the continents during the last 60 million years must have had much to do with the onset of the ice ages as well. The southward drift of Antarctica to cover the South Pole was one factor, as was the drift of North America and Greenland to their present positions, for these latter movements effectively created a landlocked sea, the Arctic Ocean, covering the North Pole. Isolated from any warm ocean currents, the Arctic Ocean soon became covered by reflective pack ice and further cooled the earth. The separation of the continents and the creation of ice caps may have put the earth just at the threshold of glacial formation, and precession-induced temperature regimes may have pushed our planet into the long, Ice Age winter.

5

By the dawn of the Ice Age, the animals and plants of our world had taken on an increasingly modern appearance. In the seas, the benthos became dominated by clams, snails, and echinoderms still present today; bony fish and sharks of that 2-to-3-million-year-old world would also seem familiar and little different from those of our world. The great forests and grasslands were largely made up of still-living species. Birds, amphibians, reptiles, and invertebrate land animals would also be relatively familiar. The land-mammal faunas, however, contained many fabulous and storied creatures that are now but eroding bones.

The modernization of land animals was in many respects attributable to changes in the world's floras. The long cooling of the Cenozoic Era created the vast grasslands, and these, in turn, dictated the evolution of new types of herbivores and carnivores. Grasslands support a greater variety of *large* animals than forests or mixed woodlands do. Prairies, because of their open nature, offer far fewer hiding places than forests;

animals of the grasslands had to rely increasingly on flight rather than camouflage and hiding to escape predators. In these environments large size became advantageous; large animals have a better chance of seeing advancing predators, while large size is necessary for rapid running. The evolution of horses well illustrates this change. Horses first evolved in Early Cenozoic times and were initially small forest dwellers. As the forests dwindled, however, and were replaced by prairies, the horses adapted to the new conditions by evolving larger size, reducing their ancestral, five-toed feet to but a single strong hoof, and drastically altering the nature of their teeth and jaws to accommodate a diet change from soft leaves to hard grass.

Although it can be deduced that the spread of grasslands favored new types of animals, it also has been argued that the evolution of the ungulate herbivores equally aided the rise of grass as one of the dominant plants of this earth. Grass plants have tenacious roots and can be cropped almost to ground level without being killed off. The same cannot be said of most other vegetation. Heavy grazing will kill most trees and shrubs, but not grass; as the mammalian herbivores increased in numbers during the Cenozoic Era, a synergistic effect took place; the spread of grasslands due to climatic cooling favored the evolution of herbivores adapted to eating grass. At the same time, these herbivores disrupted and eliminated most nongrass plants they grazed upon, thereby further helping the spread of the prairies.

The diversity of grass-eating mammals present on the earth some 2 to 3 million years ago was impressive. The odd-toed and even-toed ungulates were dominant. The great success of the horned ruminants, which include cows and deer, may be due in large part to their highly evolved digestive system. The four-chambered stomach of a cow is far more efficient at deriving nutrition from poor grass than is a horse's stomach.

Elephants were another successful group of herbivores. Today there are but two species of elephants left in the world; during the Late Cenozoic and during the Ice Age, however, elephants were both speciose and common in most parts of the world. Great mammoths and mastodons lived both in forests and on the grasslands, and thrived both in heat and extreme cold. They were rivaled in size by giant camels and by enormous ground sloths. An equally impressive diversification of small herbivores also was taking place, especially among the rodents and rabbits, both small enough to live in the grass or within burrows. Both of these latter groups fed largely on grass and other small plant seeds and thus also benefited from the rise of grassy habitats.

The changeover to a land-animal fauna dominated by herbivores of large body size produced a change in carnivores as well; as herbivore size increased, so too did body size among the hunters. Giant dogs, cats, and bears all evolved; the large, protruding canines of the saber-toothed cats were adaptations for piercing the tough, thick hide of the large herbivores. Such was the state of the world's land animals, 2.5 million years ago, at the dawn of the Ice Ages and of the Third Event.

Geological evidence indicates that ice began to cover Antarctica about 2.5 million years ago. But unlike the prior period of ice cover, now great ice regions began to cover the North Pole as well. Ice sheets began to creep across North America, Europe, and Asia, until a mile of ice gripped as much as one-third of the area of those continents. Huge glaciers crawled out of the great north–south mountain chains as well, and the earth's climate changed rapidly. Rain forests dried, deserts became wet, sea level dropped, and species began to die. In Africa, several new species of land-dwelling primates appeared. They grew in stature, and became smarter, and developed tools—and a taste for meat. Africa, giant as it is, became too small a home. Mankind began to trek from its ancestral homeland to the four corners of the globe, and greatly changed the world in the process.

6

All three of the great mass extinctions show an eerie similarity: All began with global temperature change, coinciding with a large drop in global sea level. Near the end of the Permian Period, some 247 million years ago, climate change and a drop in sea level coincided with and perhaps initiated the first great wave of marine extinctions. They also coincided with the first pulse of protomammal extinction in South Africa's Karroo desert and predated the final, more catastrophic episode by several million years. Near the end of the Cretaceous Period, about 67 million years ago, another great drop in sea level occurred, killing off our planet's reefs and many other bottom-dwelling marine creatures as well; this particular event was accompanied by climate changes that created extinctions among land floras of the time, and perhaps some dinosaurs too. It occurred about 2 million years prior to the final extinction of the dinosaurs. Two and a half million years before the present time, the same pattern appears to have occurred. The earth's climate underwent an enormous change, caused by the onset of the Ice Age. As ice sheets began to cover North America and much of Europe, the level of the sea dropped drastically, and the oceans and land cooled. And in the process, marine and terrestrial species began to die.

One of the great misconceptions regarding the current extinction is that it is just now beginning. Another is that it has so far spared marine creatures. The latter is clearly not the case. In recent years, paleobiologist Steven Stanley has documented a great and previously unrecognized loss in diversity among western Atlantic and Caribbean mollusks during the last 2.5 million years. Stanley calculated that at least two-thirds of bivalve and gastropod mollusks living in this region of the Atlantic Ocean—a fauna originally composed of more than 3,000 species—has gone extinct. He found that the loss of 2,000 species of mollusks in this region of the world alone was concentrated about the time that sea level began to drop and global climates began to cool with the onset of the Ice Age—2.5 million years ago. Other marine species elsewhere in the world began to disappear at this time as well, but studies of this event in the seas are in their infancy in many parts of the world; the listing of the dead, first marine victims of the Third Event, has just begun.

At the same time that marine life began to die off in the seas, land creatures also suffered a first pulse of extinction. Paleontologist Elisabeth Vrba of Yale University has documented a severe extinction among African mammals 2.5 million years ago, with antelope species being particularly devastated. Paleontologists have found that the North American mammalian fauna was greatly affected as the climate cooled. About 4 to 5 million years ago, the drop in sea level resulting in the drying of the Mediterranean Sea also brought about a series of North American plant extinctions and community reorganization, accompanied by a slow changeover from a browsing to grazing assemblage among mammals as the grasslands spread. Then, between 2.5 and 1.8 million years ago, thirty-five genera of North American land mammals —about 30 percent of the total fauna—went extinct as the onset of the Ice Age and first glacial advances greatly disrupted North American climate and ecosystems. But these extinctions were but the opening act of the Third Event. Just as the great climate changes of the Late Paleozoic and Mesozoic eras reduced world diversity and created an unstable and perhaps fragile series of ecosystems, so too did the great perturbations of the Ice Ages during the last 2.5 million years hammer our world. Like a boxer pummeled by too many jabs and body blows, the earth's creatures were ripe for a fall. The knockout punch, at least for North American mammals, was delivered about 11,000 years ago: Over a period of 1,000 or 2,000 years, two-thirds of North and South America's larger mammals suddenly disappeared. Many people think it no coincidence that this great extinction coincided with the arrival of mankind in the Americas.

Chapter Nine

Overkill

I

The Wenatchee valley sits in the middle of the Pacific Northwest; it is sheltered by the nearby Cascade Mountains and hence enjoys cold winters and hot summers, making it ideal for growing the best-known crop produced in the state of Washington: apples. Wenatchee is the center of the state's apple industry, and most of the land I can see on this early spring day is covered with trees, all soon to burst into blossom in a most magnificent floral display. Looking outward across the valley, I see the particular plot of land I have come to visit, an orchard that recently yielded a rich harvest far different from its usual fruit. In 1987 workers installing a new irrigation system among the apple trees of the Richey Orchard looked in wonder as their trenching machine disgorged a large stone spear point from the rich, loamy earth. The workers ceased their digging when more of the points began to appear, and to their great credit, instead of pocketing the artifacts, they called a well-known amateur archaeologist. This man must have been spellbound when he first beheld the unexpected harvest. Six inches long and made of translucent chert, the spear points were characteristic of an ancient culture first discovered near the small town of Clovis, New Mexico. Remains of the Clovis culture have subsequently been found at many localities scattered across North America. It was a culture that transformed North and South America over ten thousand years ago, created by a people who have been the center of ongoing controversy. But the artifacts coming from the Wenatchee apple orchard were larger—much larger—than any Clovis point ever discovered. The site was sealed, and professional archaeologists were called in.

By good fortune, one of the world's leading authorities on the Clovis people—and on the extinctions that wracked the Americas during the Ice Age—was located within a three-hour drive from the site. Professor Don Grayson of the University of Washington has spent over two decades dealing with questions raised by archaeological sites aged between 12,000 and 10,000 years and was well acquainted with the Clovis culture. The owner of the Wenatchee apple orchard yielding the artifacts, a Seattle area plastic surgeon named Mack Richey, visited the University of Washington and asked if Grayson would lead a new excavation at the site. Grayson, overjoyed at the prospect of such a scientific opportunity so close to home, readily accepted, but later advised Richey that an observer from the Colville Indian tribe of eastern Washington had asked to be included in the dig, as there was the possibility that human remains as well as artifacts could be unearthed during the excavation. To Grayson's astonishment, Richey refused, stating that the inclusion of Indians at the dig would bring only trouble. To his everlasting credit in my eyes, Don Grayson thereupon refused to be associated with the dig.

A second excavation of the site eventually took place, overseen by archaeologist Peter Mehringer. But Mehringer, too, found the conditions imposed on his work to be overly confining, and left the dig. Richey finally found an archaeologist more to his liking in Dr. Michael Gramley of the Buffalo Museum. Gramley opened the site for the third time, in 1990, and uncovered a treasure trove of new material. By the end of the excavation, the total list of all artifacts recovered included fourteen fluted points, four side scrapers, four ax heads, three prismatic blades, twelve bone tools, and an assortment of stone flakes and preforms. The orchard had become one of the richest Clovis sites known from North America.

By 1992 Richey was poised to extract a great deal of money out of his orchard's earthen hole. The State of Washington offered to buy the artifacts, since Richey refused to donate them to any museum; the state offered the sum of $250,000 (even though the Smithsonian Institution valued the finds at only $40,000). Richey also agreed to sell the site to the Northwest Archaeological Society (a private group hoping to preserve the site for future scientific research) for $500,000, but he broke this deal. Ultimately he sold a small portion of the orchard to a society interested in preserving the site.

The Clovis culture is the center of one of the great controversies of modern-day archaeology, a debate that spills over into paleontology as well. A nomadic people, the Clovis people appear to have entered North America from Siberia around 12,000 years ago, or soon after the retreat of North America's great glacial cover. They found a continent empty of humans (or nearly so) but filled with great animals. Mammoths, mastodons, giant ground sloths, horses, camels, giant bears, and saber-toothed cats; the list is long. Imagine the plains of eastern Africa covered with large mammals, and you can conjure a picture of the North American continent found by the Clovis people some 12,000 years ago. But within about 1,000 to 2,000 years of their arrival, most of this game was extinct. Was it climate change, or the actions of the Clovis people that killed off the Ice Age megafauna?

2

The great Ice Age mammals have played a large part in the history of paleontology: They were key evidence in the early-nineteenth-century debate on whether anything has ever gone extinct. Today it seems so ludicrous, in the light of estimates suggesting that as many as 100 species per day are currently disappearing from the face of the earth by extinction, to consider that the very reality of extinction was long disputed. In the end, Baron Georges Cuvier and others finally convinced their doubting colleagues that such great creatures as the woolly mammoths and ground sloths were simply too large to be still holding out in some lost corner of the earth. It was also apparent in the early nineteenth century that North America had lost even more of its Ice Age fauna than had Europe, causing Darwin to lament in 1836: "It is impossible to reflect on the state of the American continent without astonishment. Formerly it must have swarmed with great monsters; now we find mere pygmies compared with the antecedent, allied races."

With the realization that a great bestiary had gone extinct, and relatively recently at that, scientists studying the phenomenon naturally began searching for a cause. Cuvier hypothesized local incursions of the sea, while his more nonsecular colleagues postulated that the world-covering flood described in the Bible would have done the job nicely. Louis Agassiz, the discoverer of the Ice Ages, assumed that the world had been completely covered in ice, not floodwater, but that the results were the same: massive extinction. But other naturalists of the time looked beyond such catastrophes, to see another potential agent of destruction: mankind.

Charles Lyell is considered the father of modern geological science; his texts on the subject influenced many of his contemporaries, including Darwin himself. Thus, when Lyell suggested that the agencies of mankind may have produced past extinctions, many naturalists of the time took note. Lyell noted prophetically: "We must at once be convinced, that the annihilation of a multitude of species has already been effected, and will continue to go on hereafter, in a still more rapid ratio, as the colonies of highly civilized nations spread themselves over unoccupied lands." Initially Lyell was unconvinced that humans and the Ice Age beasts, such as the mammoths, mastodons, and great sloths were contemporaneous. As increasing evidence showed the antiquity of humans, however, Lyell began to suspect that mankind had at least helped exterminate many of the Ice Age mammals, noting "the growing power of man may have lent its aid as the destroying cause of many Pleistocene [Ice Age] species."

The debate about the cause of the extinctions continued for a century but was largely composed of opinion and pronouncement rather than data gathering and hypothesis testing. This sad state of affairs was in no small way due to the difficulties of artifact and fossil dating, for at that time there was no reliable way of correlating or dating last-known occurrences of various fossils in the glacial sediments. In the mid-1950s, however, a powerful new tool revolutionized archaeology and Ice Age paleontology: radiocarbon, or carbon 14, dating.

By comparing the relative fractions of the relatively rare isotope of carbon, C14, to its far more abundant sister, C12, a method of actually determining the age of some organic component, such as bone or wood, became readily available. Using this new technology, many scientists began to date the last occurrence of North American Ice Age fossils. A young scientist named Paul Martin from the University of Arizona, working with newly derived radiocarbon dates, found that many of the last occurrences of now-extinct mammals from North America seemed to date from approximately the same time period, a time immediately after the retreat of the last known North American ice sheet. He also noted that unlike other extinctions from the geological past, where many diverse groups of animals and plants fell victim, only a very restricted group of animals seemed to have disappeared during the Ice Age: Almost all were large mammals. In 1963 Martin staked out a position that he maintains to this day: "Large mammals disappeared not because they lost their food supply, but because they became one." In 1967 Martin published his theory in great detail. He noted that the first known humans to have settled North America, the Clovis people, did so between 12,000 and 11,000 years ago. He also noted that by about 1,000 years after this initial colonization, most or all of the extinctions among large North American land mammals had been completed. Martin proposed that the Clovis people rapidly hunted many species of the great North American mammalian fauna to extinction. This now-famous elaboration of the earlier ideas espoused by Lyell and other nineteenth-century naturalists is known as the Overkill Hypothesis. Three decades later, Paul Martin still remains its most forceful proponent.

Martin was able to make powerful arguments supporting his thesis. He noted, for instance, that the extinctions devastated only large mammals, their predators, and the scavengers that would have been ecologically dependent on the extinct mammals. If climate or some other agent had produced the extinctions, he argued, it should have cut a much wider swath through North America's biota, yet invertebrates, small mammals, reptiles, and amphibians did not seem affected by the megamammal extinction. To Martin, only human predation could account for the observed extinction patterns. He concluded that the Clovis people first arrived in North America from Asia, passing through an ice-free corridor east of the Rocky Mountains. They found a wide land, in places newly free of the great ice sheets, a land empty of humans but filled with big game. The Clovis people are considered to have been expert hunters, their skills honed by the hardships of their long Siberian habitation and eventual trek through the cold, northern wastes. Armed with exquisitely produced stone spear points, they quickly began to decimate the great herds of mammals. With a plentiful food supply, the

Clovis people quickly increased in number and spread across the continent. Martin called this rapid spread of the Clovis people, leaving behind slaughtered populations and extinct species in its wake, a blitzkrieg. His blitzkrieg model envisions a mobile group of humans, well equipped and skilled in big-game hunting, passing through previously uninhabited continental areas and so quickly exterminating the big-game fauna that few or no kill sites are left behind.

No matter what the cause, the extinction of larger land mammals in North America was rapid and devastating. According to Martin and others, thirty-five genera, spread out over a giant continent, disappeared forever during a 1,000- to 2,000-year period.

Not all of the larger mammals of North America went extinct, for twelve genera are still extant. But all of these survivors share a curious similarity: All were late arrivals to North America, arriving by the same land bridge between Siberia and Alaska that was traveled by the Clovis people. Coming as they did from either Europe or Asia, all of these mammals had long experience with humans. The great Scandinavian paleontologist Bjorn Kurten took note of this fact several decades ago: "It is noteworthy that most of the Eurasian invaders of North America —the moose, wapiti, caribou, musk ox, grizzly bears, and so on—were able to maintain themselves, perhaps because of their long previous conditioning to man." Martin agrees. He views the survivors as more gracile and wary than those killed off, animals that are unpredictable in their movements and difficult to hunt. In his view, behemoths such as the ponderous mastodons and mammoths, gargantuan but slow ground sloths, and large camels were easy targets for the nomadic Clovis people, themselves survivors of the harsh Ice Age, a people who trekked from Asia to find themselves in a warming continent amid game that had never before seen mankind. And as the giant herbivores disappeared, a suite of great carnivores also disappeared, including a North American lion, the giant dire wolf, great bears, and perhaps most fearsome of all, the saber-toothed cats.

If true, the Overkill Hypothesis should apply to areas other than North America; there should be equivalent extinctions on other continents or islands soon after the arrival of mankind. Paul Martin argues that precisely this pattern is observable in the fossil record, and cites the extinctions of large animals in both South America and Australia as cases in point.

During the Cenozoic Era, South America was separated from Central and North America by a deep expanse of sea, and hence its fauna underwent a quite separate evolutionary history. Great and unique mammals evolved there, including strange armadillolike creatures called glyptodonts as well as the giant sloths (both of which later migrated northward and became common in North America), llamas, giant pigs, huge rodents, and some strange marsupials. In Late Cenozoic time, tectonic forces caused North and South America to be joined by a land connection, and a rapid faunal exchange occurred. Some mammals of each continent mixed in with the native fauna.

As in North America, a devastating extinction occurred among South American mammals soon after the end of the Ice Age. Forty-six genera are now known to have gone extinct sometime in the last 15,000 years, and most or all of these extinctions appear to have been completed by 10,000 years ago. And as in North America, the large-mammal extinctions occurring in South America appear to have occurred soon after the arrival of mankind. The results appear to accord well with Martin's predictions; if anything, the extinctions occurring in South America were even more devastating than those in North America.

Of all of the continents, Australia has seen perhaps the greatest loss of its megafauna. The tragedy of Australia's loss was the unique nature of the extinct animals. The Australian continent, cut off from the mainstream of Cenozoic Era mammals, became the center of marsupial, rather than placental, mammalian evolution, and it was among a wide suite of extraordinary, giant marsupials that the knife of Ice Age extinctions fell.

Humanity reached Australia much earlier than it did North or South America. Perhaps not coincidentally, the wave of extinctions assailing the Australian megafauna began earlier than it did in the Americas. Good evidence now shows that mankind reached Australia no later than about 35,000 years ago, and some archaeologists conclude that humans may have been present there as early as 50,000 years ago. Most of the larger Australian mammals were extinct by about 30,000 to 20,000 years ago.

The devastating extinction striking the Australian fauna during the last 50,000 years left only four species of large native mammals alive. Unlike North America, where a continued influx of new species arrived from Asia or South America to somewhat balance the losses, no new arrivals bolstered the disappearing Australian fauna. Thirteen genera of marsupial mammals, composing as many as forty-five species, disappeared from the continent. Many of these creatures must have been extraordinary indeed. The victims included several species of hippo-size herbivores called *Diprotodon* (whose fossils can be found in extraordinary profusion), several giant wombats, a group of deerlike marsupials, and several giant kangaroos; the largest of these kangaroos was ten feet tall and weighed as much as 500 pounds. Large koalas were also present; the modern-day koala is the sole survivor of a once-diverse family. Marsupial carnivores were lost as well, including a large lionlike creature and a doglike equivalent. (The Australians refer to the former as the giant killer opossum.) In more recent times, a third predator, a catlike equivalent found on islands, has also disappeared. Several extraordinary reptiles also became victims, including a giant monitor lizard the size of a large horse, a giant land tortoise, a giant snake, as well as several species of large flightless birds, among others. The larger creatures that did survive were those capable of speed or nocturnal. Most of the Australian megafauna was ponderous, and possessed brains far smaller than many placental mammals of similar size; the Australian larger animals were apparently a relatively dim-witted bunch.

Finally, what of the areas where mankind has a long history, such as Africa, Asia, and Europe? Since humanity has long inhabited these regions, the Overkill Hypothesis would predict that fewer extinctions would have occurred than in the Americas or Australia, since the hunters and hunted would have shared many tens of thousands of years of cohabitation and coevolution together. This is, indeed, the pattern that emerges. In Africa, extinctions occurred 2.5 million years ago, but later losses, compared to other regions, were far less severe; you have only to look at any *Nature* program on PBS (or better yet, visit one of the African game reserves) to see what the great Ice Age fauna looked like and to realize how impoverished the Americas are in comparison to Africa. This is not to say that Africa and the other regions were unscathed; the mammals of northern Africa, in particular, were devastated by the climatic changes that gave rise to the Sahara. In eastern Africa, little extinction occurred, but in southern Africa significant climate changes occurring about 12,000 to 9,000 years ago were coincident with the extinction of six species of large mammals. In Europe and Asia there were also fewer extinctions than in the Americas or Australia;

the major victims were the giant mammoths, mastodons, and woolly rhinos.

Paul Martin and other adherents of the Overkill Hypothesis have amassed a tremendous amount of information and data in support of their theory. Their arguments have been powerful and skillfully presented. In a recent summary article, Martin has listed eight attributes of the Ice Age extinctions that he considers especially important in the debate:

1. *Large mammals were the primary creatures going extinct.* Mammal species with average weight of 100 pounds or more showed the highest extinctions.
2. *While the larger species were disappearing, very few extinctions occurred among small mammals.*
3. *Large mammals survived best in Africa.* The loss of large mammalian genera in North America was 73 percent; in South America, 79 percent; in Australia, 86 percent; but in Africa, only 14 percent died out during the last 100,000 years.
4. *Extinctions could be sudden.* One of the most surprising—and disturbing—features of the Ice Age extinctions was the rapidity with which entire species could be lost. Much of the debate over ancient extinctions relates to their rate over time. Unfortunately, we simply do not have the technology to discriminate, in ancient rocks, blocks of time lasting even as long as 10,000 years. For the Ice Ages, however, the powerful carbon dating techniques do allow very high time resolution. These techniques have shown that some species of large mammals may have gone completely extinct in 300 years or less.
5. *The extinctions took place at different times in different places.* Unlike the Cretaceous extinctions, where the final die-off took place simultaneously all over the earth, the Ice Age extinctions took place at different times in different places. In the Americas, they occurred about 11,000 years ago; in Australia, perhaps 30,000 years ago.
6. *The extinctions were not the results of invasions by new groups of animals (other than mankind).* It has long been thought that many extinctions take place when new, more highly evolved or adapted creatures suddenly arrive in new environments. This did not occur

in the Ice Age extinctions, for in no case can the arrival of some new fauna be linked to extinctions among the forms already living in the given region.

7. *Extinctions occurred soon after the arrival of mankind.*
8. *The archaeology of the extinctions is obscure.* One of the curious aspects of the extinctions in North and South America is that few archaeological sites yield the remains of extinct creatures; only mammoths and mastodons have been found in kill sites, while in Australia no kill sites at all have been found. Critics have often pointed to this aspect as the most powerful argument against the Overkill Hypothesis. Martin and others, however, believe that extinctions happened so quickly that there is only a small window of time containing sites with evidence that humans were responsible.

The Overkill Hypothesis has generated enormous controversy, for it is a highly charged and emotional issue. Who wants to believe that the first Americans were hunters of such skill that they could destroy thirty-five genera (and many more species) of large mammals in a single millennium after their arrival in North America? Yet it is an issue that must be clarified. If mankind could so quickly destroy the majority of the world's big game with a primitive Stone Age technology, what hope have the world's creatures in the face of our far more advanced technology? If the Overkill Hypothesis is false and the extinctions can be shown to have been caused by natural forces, such as the extensive climate change coming with the end of the Ice Ages, we face an even more disturbing set of implications. No one disputes that the extinctions took place, or that they occurred very quickly. But if such massive extinctions can take place because of climate perturbations, the world's remaining biota is in very grave danger in light of what our species is currently doing to the global atmosphere. I fervently hope that Paul Martin and the other advocates of Overkill are correct; the alternative paints a horrifying picture for the next millennium in the earth's history, a time when the Third Event will be in full swing. Perhaps we can teach ourselves to stop killing animals and thus stave off the worst potential ravages of a mass extinction. But can we change the weather? Can we stop global warming?

3

Paleontologists are not the only time travelers. Right around the corner from my office in the Burke Museum at the University of Washington resides another time traveler, an anthropologist who often journeys back into time, but to periods far more recent than my Mesozoic haunts. He journeys back to North America near the end of the Ice Ages, when a warming land was soon to be first viewed by man. Here is a description of that land by Don Grayson, the same man who politely yet resolutely declined to associate with a capitalistic orchard owner so ready to exploit the Clovis site at Wenatchee:

> What would I see in North America at, say, 15,000 years ago, when at least the bulk of paintings and engravings at Lascaux [the famous cave paintings in France] had been completed? I would see massive glaciers in the mountains of the west and covering much of what is now Canada. I would see vast lakes between the Sierra Nevada and Rocky Mountains, covering what is now mostly sagebrush desert. I would see huge expanses of pinyon, juniper, and oak woodland in the Southwest, covering land that now harbors saguaro cactus, mesquite, and creosote bush. I would see spruce woodland in the Great Plains, tundra near the glacial front in the Great Lakes region, and spruce forest to the south of that. I would see woolly mammoth, mastodon, horses, antelope with four horns, camels, and a series of mammals closely related to llamas. I would see mountain goats living in the Grand Canyon and musk ox living in Utah. I would see beaver the size of black bears, capybaras the size of Newfoundland dogs, and in the Southwest, large, lumbering sloths the size of giraffes. To hunt them I would see lions, cheetahs, and two different kinds of saber-toothed cats, and I might see giant short-faced bear. "Lions and tigers and bears, oh my!" Dorothy said about the land of Oz. If she had only said "lions and cheetahs and bears" she could have been describing what I would have seen in North America some 15,000 years ago.

What time-traveler Grayson does not see, however, on his voyages back to North America before the end of the Ice Ages is much evidence of mankind. There is currently great debate about the timing of people's arrival in the Americas, with some evidence suggesting that ancient people predated the arrival of the Clovis by as much as 5,000 years. But if such a people did live in North America 15,000 years ago or more, at the height of the last glaciation, they left very little record of their presence. The Clovis people, arriving about 12,000 years ago, however, left an indelible mark on the continent. Don Grayson has surely voyaged back among the Clovis. But in his time traveling back to that era, Grayson has arrived at conclusions far different from those of Paul Martin and the Overkill adherents: Don Grayson has emerged as the most visible and eloquent of those arguing *against* a people-produced extinction of the large, Ice Age mammals.

Much of the controversy about Overkill comes from emotion rather than science; it is unpalatable for many groups of peoples, and especially Native Americans, to consider that the first Americans may have perpetrated slaughter on such an unprecedented scale. Grayson has superbly defined the various scientific questions and cast them as testable hypotheses. Even Paul Martin acknowledges that Grayson has markedly improved the entire scientific issue by his careful restructuring of the debate.

Grayson's major critique of the Overkill Hypothesis stems from what he sees as problems in chronology. Before the advent of carbon 14 dating techniques, the timing of the various large-animal extinctions around the world was quite problematical. But as increasing numbers of radiocarbon dates began to accumulate in the literature, patterns in the extinctions began to be perceived. By the time of Paul Martin's seminal 1967 paper on Overkill, a large amount of dated material was available; this data set convinced many scientists that the main wave of extinctions had culminated about 11,000 years ago. After contemplating the chronological data at hand, Martin made two assumptions: First, since no fossil remains of extinct mammals were found in any archaeological site age 10,000 years or younger, he assumed that *all* of the extinctions had been completed by that date. Second, he assumed that whatever had caused the extinctions of some of the animals had caused the extinction of all; and since he had last appearance dates (the radiocarbon date for the last-known existing individual of a now-extinct mammal species) ranging between 12,000 and 10,000 years ago for thirteen of the thirty-four genera known to have gone extinct, he concluded that all of the genera went extinct in this interval.

According to Don Grayson, who has written several perceptive articles about the scientific history and philosophy of the Overkill issue, Martin completely restructured the issue through the force of his arguments:

> Martin's 1967 paper is the most influential analysis of the North American extinctions ever written and fundamentally altered the way in which people thought about those extinctions. Martin's arguments regarding the chronology of the extinctions may have fit his own ideas concerning the causes of those extinctions. But the dates he provided also aligned the extinctions with increasingly secure evidence for major climatic change at this time. As a result, his chronology fit almost everyone's ideas on those causes as well. That is, Martin's position concerning the timing of the extinctions seems to have been widely accepted not so much because of the strength of radiocarbon chronology that was then available, but because of the kinds of explanations felt most likely to account for those extinctions. It is thus no surprise that most scientists quickly abandoned the notion that the extinctions may have been spread out over thousands of years. In its place, they adopted the position that all or virtually all of the losses had occurred between 12,000 and 10,000 years ago. Unfortunately, analyses of the radiocarbon dates now available for these animals do not provide much support for this belief.

According to Grayson, many of the radiocarbon dates Martin used have now been shown to have been biased by either poor material or techniques inferior to those used today. As in any branch of science, laboratory techniques and methods usually improve as better machines and analytical methods become available. The methodology of C14 dating is no exception; tremendous strides in technique during the last two decades have both pushed back the window of resolution and made age determinations far more accurate. Grayson suspects that many of the dates utilized over two decades ago may be flawed; he has concluded that the bulk of the extinctions may not have taken place during the critical interval between 12,000 and 10,000 years ago, but that many species were already extinct by the time the Clovis people arrived in North America. The same argument can be applied to South America. The Clovis people may have wiped out some species, but not the entire batch of thirty-five genera as proposed in the original Overkill model. Another telling criticism of the Overkill scenario is that so few mammal skeletons are found in "kill sites," places where the fossils of extinct species are associated with either human artifacts used to kill the animal or evidence of human activity, such as bone butchering. To date, kill sites remain rare, and very few of the total number of extinct mammal species can be found at such sites. If early people were slaughtering animals in such numbers that rapid extinction ensued, they must have been doing it very covertly—or, as Paul Martin has argued, it happened during a very narrow window of time.

The major alternative to the Overkill scenario is that the extinctions were the result of rapid and profound climate changes following the retreat of the glaciers. There is no doubt that great changes in climate *were* taking place around the globe while the mammalian extinctions were occurring. But did these climate changes kill anything? The major criticism of the climate change hypothesis is that many such climate changes occurred during the numerous glacial cycles of the last 2 million years, but very few mass extinctions have occurred during that time. It cannot be demonstrated that climate conditions at the end of the last glacial period were any more severe than those of fifteen or twenty other interglacial periods, yet the extinction since the retreat of the last ice sheets has been far more devastating than any other during the last 2 million years.

The most detailed climate-induced model for extinction does not rely on sudden temperature or moisture change per se. Developed by Russel Graham and Ernest Lundelius, this model suggests that the cool but equable conditions known to have characterized the late glacial period changed to warmer but more extreme temperature regimes following the glacial retreats. These climate changes affected North America's various plant communities, causing them to become less diverse and thus less able to support a diverse assemblage of mammals. Small mammals migrated to new regions, but large mammals, requiring more food, died out. There is no doubt that the end of the Ice Age was accompanied by sudden and drastic changes in temperature, and that a dramatic change in plant communities and their distributions across the North American continent occurred soon after. But the idea that all the larger mammals were unable to migrate out of harm's way seems unlikely; we know that many large African mammals are perfectly capable of making long treks in search of seasonal food sources or water. Climate change alone seems unlikely to have killed off thirty-five genera of North American mammals so rapidly, in 2,000 to 5,000 years.

The issue of large mammal extinctions has been the source of a great scientific debate for more than a century and a half, and if anything, the debate has only intensified over the past two decades. No single explanation seems reasonable at the present time, and somehow, knowing the history of other great extinctions in the earth's long past, this does not surprise me. For all of our experience, the mechanism of extinction still remains mysterious. A multicausal explanation will surely be necessary to unravel the great loss of wondrous creatures living so recently on our earth. Climate change surely put great stress on the earth's ecosystems. But the ravages of hungry people were surely involved as well in the destruction of many species now extinct.

4

The Clovis site found in Wenatchee, Washington, was deposited along the edge of one of the world's great rivers: the Columbia. Starting high in the Canadian Rockies, the Columbia gradually swells in size until it reaches the State of Washington. The rich record of artifacts recovered so far represents one of the largest caches of material yet recovered from any Clovis site and might be multiyear habitation (or it could simply be a cache used once). Perhaps the Clovis people used the Columbia for transportation, canoeing up- or downstream in pursuit of wildlife, raw materials, or trade. Or perhaps they camped there because of the rich abundance of game on the neighboring grasslands and valleys. Whatever the reason, the giant river served as their home.

The Clovis people appeared to have undertaken great migrations; often the artifacts found are made of stone or other material that must have originated hundreds or even thousands of miles from its final resting site. Between 11,000 and 9,000 years ago there seems to have been a small number of Paleo-Indian cultures—the Clovis culture being one of these. And then, about 9,000 years ago, a great diversification of cultures began, culminating in the many tribes of Indians found in North America today. In essence, the Clovis people themselves became extinct, as they evolved into the great diversity of Native Americans.

As the Clovis people spread across America and diversified into the countless North American and then South American tribes, great climatic, geographic, and vegetational changes were taking place in North America. Large regions that had been under ice or covered by giant lakes emerged as dry land, at first barren, but eventually colonized by rapidly changing plant communities. The surviving mammals flourished. White-tailed deer spread across North America, while mule deer ranged through the high plains and western mountains. Huge herds of pronghorn antelope spread across the western prairies, sharing the bountiful grass with herds of plains bison and elk. Prairie dog colonies covered great expanses of territory, while mountain lions, great bears, and packs of wolves vied for food among the abundant herbivores of the west. In the east of the great continent, an even larger bison lived in the vast forests, along with herds of elk and giant moose. Great flights of waterfowl and other birds blackened the skies. It was a continent rich in animal life.

This long history occupies my mind as I rest on a ledge of gritty strata overlooking the mighty Columbia River, 100 miles downstream of the Wenatchee Clovis site. I am on a lunch break, and gratefully basking in a warming spring sun after a morning's collecting among fossiliferous rocks. In this region of Washington State, the Columbia River is lined by outcrops of million-year-old strata known as the Ringold formation. It contains a rich assemblage of fossil vertebrate skeletons, and earlier in the morning I had uncovered one of the most beautiful fossils I have ever seen: As I smashed away at a ledge of pebbly conglomerate, a large block split away from the cliff beside me. Amid the yellow and orange rocks, tinted by rich iron concentrates, a great chunk of elephant tusk suddenly lay revealed. It created a moment of disorientation in my mind: This fragment of ancient ivory, so clearly from a mammoth or mastodon, seemed an impossible object to be suddenly revealed by North American sunshine. I was too obviously in my home state, on a continent familiar from birth: There are no elephants here. And thus will the conscious mind, if given the chance, reject the notion of extinction. All of my training tells me that great herds of giant elephants long haunted this continent, but although I *know* this fact to be true, my initial reaction at uncovering direct evidence was disbelief. Yet great elephants *were* here, and they were seen, hunted, and killed by Washington State residents who lived here before me, not so many thousands of years ago. Perhaps the Clovis hunters scouted the great herds of elephants, camel, and deer from this same strategic spot on which I now sit, perched high above the Columbia. There is no doubt that the Clovis people hunted and killed North America's great elephants, for mammoth kill sites are known; whatever doubt exists about other prey of the Clovis, there is none concerning the great elephants. And if they could kill the elephants, perhaps the most dangerous of game, why not smaller, easier species as well?

Looking out over this broad plain, it is hard not to speculate about the loss of North America's great Ice Age mammals. Like so much else related to the past, there is a strong possibility that we will never know if the Overkill Hypothesis, as so strongly advocated by Paul Martin and his supporters, is indeed the primary reason that the plain below me no longer supports the fabled beasts. But it is not only the mammoths, mastodons, ground sloths, and camels that can no longer be found on the rich grassland around me and elsewhere on the North American continent; even the great herds of surviving mammals are now far less numerous than at any time since the Ice Ages. Even if Overkill is eventually abandoned as the major cause of Ice Age extinctions, the concept still carries tragic validity for times approaching—and overlapping—with the present day. Today the plain stretching below me is devoid of game. Much has been hunted, much displaced by the sprawling farms dotting the landscape. Even the giant river running through eastern Washington offers mute testimony to Overkill: Once the home of untold numbers of salmon, the once-mighty Columbia, overexploited and overfished, is now an empty series of dammed lakes, and its salmon runs are a thing of the past. The river below me does not flow by; it sits largely motionless, still as death. And in the distance an even more ominous reminder of mankind's presence sits, squat and menacing. This section of the Columbia is the site of the Hanford Nuclear Reservation, where for forty years the most poisonous substance yet known in the universe—plutonium—has been spilling into the soil, the air, and the river, a product of mankind's desire to extend Overkill to his own species.

5

The loss of North America's wildlife has been documented recently in a beautifully written but terribly disturbing book, *The Endangered Kingdom,* by Roger Di Silvestro, a former senior editor of *Audubon* magazine. Di Silvestro begins his book with a trip back into time, to North America immediately prior to the arrival of the Europeans. He inventories the wildlife then present. The numbers are impressive for those of us living in the same continent five centuries later. There may have been 50 million bison on the Great Plains alone and 40 million pronghorn antelope. We are all too familiar with the destruction of the bison, and the pronghorns suffered no less. By 1900 there were fewer than 20,000 antelope on the entire continent, and perhaps no more than 500 plains bison still existed. On the East Coast, a large flightless bird called the auk lived in large numbers; today it is extinct, as is the heath hen, a bird also once common there but hunted into extinction. Di Silvestro documents a single eighteenth-century hunt in Pennsylvania, where hunters gathered from many parts of the state. They formed a circle 100 miles in diameter, with a hunter located each half mile. The hunters marched inward, killing all they found; the final tally included 41 cougars, 109 wolves, 18 bears, 111 bison, 112 foxes, 114 bobcats, 98 deer, and more than 500 smaller mammals. Throughout the continent, both game and nongame animals have been largely removed. The California condor lives only in zoos, Stellar's sea cow is extinct. The wild turkey no longer lives throughout America, wolves have dropped in numbers from more than 2 million to perhaps 2,000 in the continental United States, grizzly bears number in the hundreds at best, and cougars are largely gone.

If there is a symbol of Overkill, perhaps it is best exemplified by the passenger pigeon. As many as 5 billion of these birds lived in North America two centuries ago. Di Silvestro recounts stories of seventy birds being killed in one shotgun blast, of a billion birds being killed by hunters from a single nesting site in Michigan that was forty miles long and ten wide, of one hunter killing 5,000 pigeons in one day, of 200,000 birds being killed in a single hunt—the last great hunt—held in 1896 in Ohio. The last wild bird was shot in 1900. The passenger pigeon was extinct soon after, a victim of Overkill.

The seas are in no better shape. The coastal waters teemed with crabs and shellfish. Now, for the first time in history, the State of Washington is considering a total ban on sport salmon fishing, at least for one year, because of the small runs; the Snake River salmon is virtually extinct and belatedly is being considered as an endangered species; king crab fishing in Alaska has been essentially terminated because the stocks are gone; the great shellfish fisheries of Puget Sound have been halted because the oysters and mussels are too poisoned by industrial wastes to eat.

Paleontologist David Raup has written more about extinction than perhaps any other scientist. He has noted the obvious: The first step toward mass extinction is the reduction of individuals. North America is today in the midst of a deepening and terrible mass extinction. We have replaced perhaps a billion mammals of many species with 250 million mammals—humans—of one species. We have done it largely through Overkill.

Chapter Ten

Lost Islands

I

The sun is blinding as I emerge from the dark crater and clamber onto the upper rim of the old volcano. Hastily donning sunglasses and still puffing from the steep climb, I finally get a chance to see the view. This highest part of Diamond Head crater on the island of Oahu is surrounded on three sides by bluest ocean, whipped into whitecaps by the strong trade winds. To the south, windsurfers are playing magnificent tricks with gravity as they fly over the surf; to the west, the late-afternoon sun is quickly dropping toward the horizon. And to the north, the Miracle Mile of Waikiki beach seems a solid wall of high-rise hotels, shimmering in the golden sunlight, the sandy shore but a thin ribbon dotted with people. Climbing atop an ancient gun emplacement amid a host of other tourists, I look back, into the crater of this ancient volcano, and try to imagine it as it once was, erupting great gouts of lava and smoke into the blue sky. But the last eruption of this eroded cone, whose outline may form the most famous silhouette in the world, ended more than 2 million years ago, before the glaciers began to cover the earth and when mankind was still a future evolutionary dream. Every year the island of Oahu grows smaller as the forces of erosion chip away at its rocky foundation, for no new mountain building or island growth rebuilds or sustains it. The land of Oahu once sat over a fountain of upwelling magma, a lithic manna that caused the island to grow over the long centuries. But the forces of plate tectonics have slowly dragged Oahu away from the source; over 200 miles to the southeast, the Big Island of Hawaii now sits over this conduit to the earth's mantle.

The Hawaiian Islands are but the most recent constructions of this hot spot, a term used by geologists to describe a great plume of magma rising upward from the mantle region of the earth's deep interior. This particular hot spot has been active since late in the Mesozoic Era. Over a cycle lasting about 10 million years, new islands are formed, sit for a time above the sea, and are then utterly destroyed by erosion. A long chain of sunken, dead sea mountains trails off to the northwest of the Hawaiian Island chain, each a former island, each with a long history of bright sunny days among forest and sandy beaches. Others, still emergent, will join this chain of the dead. Niihau, the most westerly of the current Hawaiian Islands, will be the next to disappear beneath the sea, followed by Kauai and then Oahu. Kauai, already nearly 6 million years old, is clearly in old age. Eventually even the million-year-old Big Island

PHOTO BY THE AUTHOR.

Diamond Head crater, Oahu.

itself will slide past the active hot spot. Then its life-giving volcanic furnaces will still and its size will begin to decrease.

For all of their beauty, the Hawaiian Islands somehow don't seem exotic to me; their familiarity as just another state of the United States seems to exclude them from the list of truly wild places. They are also very crowded pieces of real estate, with a million full-time human residents and another 4 million tourists arriving each year. This statistic seems totally believable based on the airplane traffic alone. Standing atop Diamond Head crater, I am almost beneath the glide path into the Honolulu airport, watching an endless stream of jumbo jets arriving with a fresh batch of paradise seekers. But before airplanes shrank the world, making Hawaii and many other formerly remote places suddenly accessible in the process, this string of volcanic islands was very inaccessible; in fact, no other parcel of island real estate on the earth is as remote as the Hawaiian Islands, for no other archipelago is more isolated from continental landmasses. North America, the continent nearest to Hawaii, is over 2,500 miles away. Because of this isolation, the plants and animals ultimately arriving on these shores did so only by beating enormous odds.

Each of the native plant and animal species now present on Hawaii either arrived here after a chance voyage or evolved from some species that did. The affinities of the fauna and flora suggest that these chance arrivals came from many directions: The South Pacific, the Indo-Pacific, and the Americas all have contributed species. Tiny seeds may have alighted following great tempests, or arrived entangled in the feathers or stomachs of visiting birds; insects probably arrived in similar fashion, or washed ashore on drifted logs after many weeks or months at sea. Oceanic seabirds probably were the first species to visit the newly emergent islands, only later to be followed by terrestrial birds, flying here with the help of the winds, perhaps.

However they arrived, new forms must have arrived very, very rarely. Scientists have concluded that the entire Hawaiian biota, comprised of many thousands of species, has evolved from about 1,000 ancestral, nonnative species; only a thousand times during 70 million years did some seed or creature arrive on the islands, take life and grow, and then propagate. Those species that did manage to cross the thousands of miles of open ocean arrived into an ecological vacuum, a place of volcanic soil, plentiful rainfall, but virtually no animal life. Evolution took these few voyagers, plant and animal alike, and transformed them over the millions of years. As new islands formed in the chain, they were colonized by immigrants from the older, nearby islands; as each of these islands, in turn, ultimately eroded and finally sank beneath the sea, it must have taken great numbers of species into extinction with it.

For 70 million years, the forces of chance arrival and evolution, proceeding in a nearly isolated system, worked their miracles, ultimately producing fauna and flora unique on the face of the earth. No other known island group contains a larger number of endemic species, composed of organisms found nowhere else on the planet. The final tally is impressive: about 1,400 species of plants, 8,000 insect species, 1,000 land snails, and over 100 species of birds are known to have evolved, as well as many other species belonging to other invertebrate groups. This total is itself artificially low, for scientists have just begun to study many groups of Hawaiian animals and plants. Some specialists believe science has documented only perhaps half of the existing invertebrate species once or still living on the Hawaiian Islands.

The Hawaiian archipelago is over 1,500 miles long. Most of the islands to the west, such as Midway and Laysan, are atolls or tiny rocky crags containing little terrestrial life. Most of the land area and species of Hawaii are concentrated in the eight main islands. Each of these islands is unique in many ways, offering characteristic topography, climate, and substrata. Most are craggy, with rapid elevation changes over short distances. Because the Hawaiian Islands are in the trade-wind latitudes, each island has a characteristic windward and leeward side, producing very different rainfall patterns. This great geographic subdivision and diversity of environments is largely responsible for the high rates of evolution and diversification that transformed the few arriving biotic colonists into the huge numbers of individual species found today. It also produced a large number of different plant and animal communities, each adapted to the particular conditions of sun, nutrients, and rainfall found in the many different habitats. Botanists have identified between 86 and 152 native plant communities on the various islands.

Before humans arrived on the islands, forests were the natural vegetation cover there; only the high alpine zones and the driest parts of the leeward lowlands bore communities dominated by shrubs and grasses rather than trees. The windward lowlands on each of the islands were apparently covered by richly diverse rain forests, which exist today only as small pockets in steep or inaccessible valleys.

The Hawaiian biotic communities had no terrestrial vertebrates other than birds, and this fact greatly affected the course of evolution on the islands. No reptiles or amphibians ever won a sweepstakes ticket to sunny Hawaii over the long millions of years; no floating palm log ever washed onto Hawaii's shores with a pregnant snake or lizard. Similarly, the only mammals native to the islands are the Hawaiian monk seal, which spends virtually all of its time in the sea, and a single species of bat. Birds were the largest vertebrate creatures on the islands. In the absence of large terrestrial predators, a variety of flightless birds evolved from winged ancestors. In a similar fashion armies of unique insects and land snails arose, developing body plans and ways of life not possible in reptilian- and mammalian-dominated ecosystems. The Hawaiian Islands were hotbeds of evolutionary change, greenhouses packed with exotic species. The Hawaiian fauna was diverse, abundant, unique. And then mankind arrived.

The Hawaiian Islands now have the dubious distinction of containing the highest number of endangered and threatened species of any state in America. But the endangered species may be the lucky ones, for they still have some lease on life, however tenuous. Many other Hawaiian organisms have not been so fortunate. We are just beginning to learn how many species have gone extinct since the arrival of mankind in these islands.

The rapidly setting sun reminds me that it is time to hike out of the crater; long shadows are already stealing across the scrub-covered walls, and the dark tropical night is not far behind. I take a last look out to sea and discern a small catamaran sailing toward the shore. It conjures up images of the great navigators arriving here nearly two millennia ago, the first humans to land in this fecund land, washing ashore surely hungry and thirsty, perhaps at hope's end, arriving by chance like current-borne seeds following a long ocean voyage. Like many chance arrivals before them, the first Hawaiian people settled on the fertile plains, and prospered. But unlike other arriving creatures, which soon produced evolutionary descendants, these new immigrants were far-traveled seeds of destruction.

2

The first humans to arrive on the shores of the Hawaiian Islands did so long ago. The first contact was long thought to have occurred about A.D. 800, but more recent archaeological investigation now shows that the first humans sailed here closer to the year A.D. 300, or 400 at the latest. They may have arrived from the Marquesas, a group of islands 2,400 miles to the southwest of Hawaii. This identification is at best a guess, based on circumstantial evidence such as similarities in language, body type, and shared agricultural plants; no written record that could provide clues to this mystery exists from this long-ago event. Some archaeologists believe that at least two different South Pacific island groups contributed people to the Hawaiian Islands; they often mention the Society Islands, which are near Tahiti, as the second possibility. Whatever the source or sources, however, the people who arrived here in their small outrigger boats did so only after long and perilous voyages. Many failed voyages also must have taken place, trips ending in dehydration, starvation, and hideous death in the seemingly endless, empty Pacific, for the chances of arriving at an island group as small as Hawaii after traveling such a great distance are remote.

We know that the earliest human arrivals brought immigrant plants and animals with them, most intentionally, but some as inadvertent stowaways. At least thirty-two plant species arrived with the early Polynesian settlers, including food-crop plants such as bananas, taro, sugarcane, yams, sweet potatoes, and coconut. Other nonfood but essential plants also imported included gourds, bamboo, and plants useful in the manufacture of tapa cloth, dyes, and oil. All of these plants were introduced into the Hawaiian ecosystems, and soon began to grow in the wild.

The Polynesians also introduced several animals. Food animals included pigs, dogs, and chickens, while several stowaways on the sailing boats also reached shore, such as the Polynesian rat, seven species of lizards, and a few land snails. Many of these alien species were benign and had no impact on the native Hawaiian ecosystems. Others, however, were far more lethal. Some of the pigs and dogs brought for food escaped the pen and leash and began to breed in the wild. These feral pigs and dogs surely began to prey on native flora and fauna, while the pigs became major agents of alien plant transport, by carrying seeds in their fur and feces. But by far the most destructive of the introduced animals was the rat, which immediately began to wreak havoc on the native bird and invertebrate populations.

By the sixth century A.D., the Polynesians had created permanent settlements on all of the main Hawaiian islands. According to the leading expert on Hawaiian archaeology, Professor Patrick Kirch of the University of California, population numbers remained low for a long period following the first arrivals; he estimates that no more than 1,000 people lived on the islands as late as two or three centuries following the initial landings. Most settlements were constructed in the rich lowland regions, areas where abundant rainfall and fertile soil allowed farming.

Human population size in the Hawaiian Islands remained relatively low until about A.D. 1100. After that, however, the number of humans on the islands began to increase rapidly, doubling each century by some estimates. The burgeoning populations required ever more food, and agriculture systems quickly expanded in size and scope to cope with the demands. The human population peaked at approximately 150,000 to 200,000 people about A.D. 1650, according to Kirch. By this time all of the favorable agricultural areas were already in cultivation and the Hawaiians had to expand their efforts into marginal environments. They did so by developing a sophisticated method of dryland irrigation, which opened up the dry scrublands to cultivation. Kirch has estimated that during the population peak, the density of humans in the fertile valleys may have been as high as 750 per square mile.

Other archaeologists have suggested that the Hawaiian population at and immediately after this time may have been much higher; some suggest a maximum population size of between 800,000 and 1 million people. Regardless, it appears that about this time, the carrying capacity of the Hawaiian Islands was reached and exceeded; the islands just could not support so many people. Further expansion of farming to very marginal soils and habitats and the use of sophisticated agricultural techniques allowed only a stabilization of population numbers, not further growth.

During their rapid, post-1100 growth period, the Hawaiian Islanders must have greatly altered native vegetation on all of the islands. Archaeological and paleobotanical studies indicate that the lowland forests were almost completely replaced by taro ponds, gardens, habitation, and introduced tree species. Clearing of vegetation was accomplished by manual tree removal and burning. During the earliest periods of human occupation, land cleared for fields was used once and then abandoned. This slash-and-burn agriculture and field abandonment exposed many formerly vegetated areas to severe erosion and soil denudation. As the human population increased, however, and new land areas became scarce, older fields had to be cultivated year after year. While this reduced erosion, the practice exhausted the soil in many areas.

By the mid-1700s the human population of the Hawaiian Islands had either stabilized or was declining slightly; an equilibrium between the number of humans and the land's ability to yield the necessary crops had been reached. The great distance between Hawaii and any other human populations had also effectively isolated the Hawaiian culture, and if not exactly an idyllic paradise in regard to the Rights of Man (the political system was a highly constrained caste system with slavery and ritual human sacrifice), it was probably not the worst of times or places to be human. It would have been interesting to see what this culture would eventually have evolved into and accomplished in the process if left alone. But that was not to be. In the latter part of the eighteenth century, after a millennium and a half of isolation, the Hawaiians had the misfortune to be "discovered" by European civilization and European diseases.

3

Waimea means "bloody water" in Hawaiian, and the name seems sadly, symbolically appropriate for the small town on the southern coast of Kauai, located near the mouth of a river disgorging tons of red sediment into the sea. Today the town of Waimea is a dilapidated assemblage of small houses, tourist traps, and fast-food franchises that identifies it as being in Anywhere, U.S.A. In the center of town, far from the sea and next to the busy coastal road, stands a bronze statue of Captain James Cook, R.N. The statue is tarnished and smeared with a bit of graffiti, and part of it has been broken away. A small plaque states only that the statue is a copy of the original, which is found in Whitby, England, and commemorates one of England's greatest navigators. Some distance from this statue, which surely seems curiously out of place and context to more than a few who pass by, stands a pillar of basaltic rock, partially hidden in the shade of several tall trees. By walking around this second monument, another, smaller plaque can be found. It states that the monument was built in 1928 by the people of Kauai, to commemorate the discovery of the Hawaiian Islands by Captain James Cook on January 20, 1778.

Whatever I was expecting to see commemorating the first landing of Europeans in the Hawaiian Islands more than two centuries ago, this was not it. I drove to the beach where the actual landing took place, expecting to see some further comment or monument, and a monument to this event I did indeed find. Driving down a dusty road, past ramshackle huts and a decaying store identified as Bucky's Liquor and T.V., Inc., I arrived at a small beachside park. Rusting car bodies made up the north wall of the park, while squatters evidently had moved in along the beachfront. Unlike the other beaches I had seen on Kauai, which were mainly pristine (perhaps due to the fact that most fronted expensive hotels), the Waimea beach was dirty brown in color from the riverborne sediment, and covered with plastic bottles, sacks, and broken glass. It was from this same beach, two centuries ago, that the Hawaiians cheered the arrival of the great, three-masted sailing ship, arriving so dramatically into the broad bay near Waimea. Who but the gods could have built such a wonder? It was as deities that the white men were welcomed. Looking down the beach, I wondered where the first of Cook's rowboats landed. A brisk breeze blew inward from the sea, carrying a plastic grocery bag in its embrace. I brought my camera, but this beach did not deserve a picture, or honor, or memory. The piles of filth and garbage now lining the beach, some surely left by descendants of those Hawaiians who first welcomed Cook, are testimony enough.

The arrival of Cook in 1778 set off a wave of death among the Hawaiians and extinctions among the native plants and animals. His ship, and the many that arrived soon after, brought a new group of species to the islands: the microbial species producing syphilis, tuberculosis, influenza, measles, and other diseases to which the native Hawaiians had no immunity. Within twenty-five years of this first contact, the population of the Hawaiian Islands had plummeted by half; within a century of Cook's arrival, only one-twentieth of the precontact population existed.

With the catastrophic depopulation of the Hawaiian Islands during the nineteenth century, many of the great agricultural fields and the villages they supported ceased to exist. The rich, newly emptied farmlands did not escape notice in the early European period. The new human immigrants seized the lands and began to grow new types of crops. By 1840 over 100 non-Hawaiian species of food and other marketable plants had been introduced to the rich, lush islands. Hawaii became an important stopping point for whaling ships and other traders, all carrying large crews requiring fresh meat, fruit, and vegetables. The whaling ships also required large quantities of firewood for their rendering boilers. For fifty years this trade persisted, and by its end the last lowland dry forests not already razed by early Hawaiian agricultural practices had been logged. At the same time a profitable trade for a tree known as sandalwood commenced, virtually stripping higher-elevation forests of this species and disrupting the surrounding forests in the process. A final forest product exported during the nineteenth century was pulu, the hairs from giant tree ferns. Once abundant in the islands' rain forests, these ancient trees, relics of the Mesozoic Era, were largely removed by logging. The Europeans introduced metal tools and improved methods of logging and wood manufacture; they cleared new roads and established logging camps in topographies and elevations unavailable to the Hawaiians with their nonmetal technologies. In doing so they completed the forest destruction and alteration begun by the Hawaiians. But an unforeseen result of the intensive logging was soil erosion and changes in the original water courses and drainage patterns; these effects became even more destructive than the logging.

Vast numbers of alien plant (and animal) species continued to be introduced to Hawaii during the late nineteenth and throughout the twentieth centuries, continuing to the present day. Where species introductions prior to man's arrival were occurring at an average rate of one every 70,000 years, between ten and a hundred new species of insects *alone* are being introduced to the Hawaiian Islands each year. Local foresters considered that Hawaiian tree species were inferior to exotics in wood quality and growth rates, and hence began tree farming using alien tree species. Fast-growing tree species of no commercial value were also imported, the better to establish watersheds and aid in reducing erosion in logged areas and impacted watersheds, for by this time vast numbers of feral cattle, pigs, sheep, and goats had also caused widespread erosional damage to many sensitive areas. In this fashion banyan, eucalyptus, and paperbark trees were planted in great numbers to help reestablish forests and reduce soil loss through erosion. All of these species were chosen because of their rapid growth, and all quickly spread into the native forests and began to displace native communities of slower-growing species.

The problem of nonnative plants in Hawaii is now extremely serious. About 1,000 to 1,500 species of plants made up the native flora prior to European contact, with about 90 percent of these found only in the Hawaiian Islands. Since Captain Cook's arrival, another 4,600 nonnative plants have been introduced to Hawaii. Most of the alien species posing the greatest threats to native Hawaiian vegetation have been introduced in the twentieth century; almost all are weeds.

Botanists generally define weeds as plants growing out of place; by this definition all of the nonnative plants now found in Hawaii would have to be considered weeds. However, agricultural crops are usually excluded from the term, as are most forest tree species; the vast fields of pineapple and sugarcane, so important to Hawaii's economy, are covered with alien species no one calls weeds. Weeds are usually considered nuisances, since they take over habitat favorable to native species and often displace or kill off the original vegetation in the process. According to botanist Clifford Smith of the University of Hawaii, most alien weeds have entered into Hawaiian floral ecosystems first in lowland ecosystems, usually following habitat disturbance, such as agricultural clearing, fire, or urbanization. Lowland weeds get transported to higher elevations in the feces of animals, such as feral pigs, and through human activities.

Newly arrived weeds are disruptive through competition for space, nutrients, or water. Particularly devastating are weeds capable of growing over other plants and smothering them. In such cases plant diversity drops markedly. Eighty-six alien plant species are now recognized as posing serious threats to the survival of native Hawaiian plants. One of the worst weeds is the banana poka, a climbing vine native to Colombia that was introduced to the Hawaiian Islands in 1920 as a garden ornamental. It escaped, and now smothers even the largest trees; it is too widespread to control. Other nasties include the blackberry (introduced for its fruit, and now found on all of the islands), several species of bunchgrasses and molasses grass (introduced as ornamental grasses, but now displacing native grasses and herbs), Koster's curse (a shrub introduced as an erosion-controlling plant that smothers all other plants and has already conquered large tracts of native vegetation in the Fiji Islands as well as Hawaii), and possibly the worst of all, strawberry guava, a rapidly growing, thicket-forming tree that shades out all other plants. Strawberry guavas now form the understory of many of Hawaii's formerly pristine rain forests. These and many other weeds have certainly reduced the diversity of Hawaiian native plant communities and are contributing to plant extinctions in the islands. Today more than 750 species of native Hawaiian plants are listed as federally recognized endangered species or are candidates to be so recognized.

The introduction of nonnative plants is not all that has severely altered the Hawaiian landscape; in the past two centuries over 2,000 alien invertebrate and 81 vertebrate animals have been introduced either intentionally or unintentionally, and most have subsequently escaped into the island ecosystems. Because of these introductions, the Hawaiian biotic communities have been forever altered.

Like the introduction of alien plants, which is now creating so much havoc and impending extinction among the endemic Hawaiian flora, many alien animal species were introduced with good intentions by biologically naive people; others were stowaways first on boats and later on airplanes. Of the latter category, the worst tragedies have come from the arrival of mosquitoes and social insects such as ants, bees, and wasps.

It seems beyond belief that any tropical place could be mosquito-free. Yet that was the happy circumstance on the Hawaiian Islands prior to the arrival of the Europeans. It was not to last. During the 1820s, a sailing ship arrived from the Americas for revictualing. The ship's freshwater supply, stored in great wooden casks, had been taken aboard in Mexico. Upon arriving in Hawaii, the ship's crew rowed the casks to shore, dumped the remaining, fetid water into a small pond, and refilled their casks with fresh water. Unbeknownst to the sailors (who probably would not have cared anyway), the old cask water contained mosquito larvae; these larvae reached adulthood, escaped, and rapidly multiplied into great swarms on every island. Although annoying nuisances for humans, the arrival of mosquitoes into the Hawaiian Islands was devastating to the native bird fauna, for the mosquitoes added the one missing link necessary to infect much of the bird fauna with deadly diseases—the vector.

Alien birds introduced into Hawaii by the Europeans during the last century brought two devastating diseases with them: avian pox and avian malaria. Because there were no mosquitoes on the island, however, at first the diseases did not spread widely. With the introduction of mosquitoes, however, the situation changed rapidly. The birds of Hawaii, like the Hawaiian people following the arrival of the Europeans, were decimated by these diseases.

If the mosquitoes proved devastating to the bird populations, the introduction of ants, termites, bees, and wasps was equally disastrous and to a far larger assemblage of species.

Flowering plants cannot reproduce without the aid of pollinators; most of these are insects. In some cases plants and their particular pollinators co-evolved very particular morphologies, and if one of the pair is endangered or reduced in numbers, the other suffers as well. Entomologists studying the highly evolved mutualism between the native Hawaiian flora and its native insect pollinators have estimated that over a thousand species of Hawaiian insects were involved in pollinating. Unfortunately, alien ant species have terribly affected these delicate plant-insect relationships.

During the last century, a large, nasty ant called the big-headed ant arrived in Hawaii, probably as a stowaway, and proceeded to colonize all of the islands. Now the dominant insect in the lowlands, it seems to have killed off virtually all of the other insect fauna below about 2,000 feet in elevation, for the only other abundant insects over most of the ants' lowland range are large cockroaches, also introduced. The ants have severely reduced the diversity and abundance of native pollinators, while doing no pollinating of their own; they eat the pollen of native Hawaiian plants but do not pass it on. Happily, the big-headed ant does not live in the highlands, areas that have become the only refuges for local Hawaiian insects. Unhappily, in 1940 a second ant species, the Argentine ant, was introduced to Hawaii and in 1952 a third, called the long-legged ant, also received its immigration papers for a Hawaiian existence. The former thrives in upland areas, and thus takes over where the big-headed ant left off; the latter is an excellent tree climber and robs most trees of pollen, thereby reducing their chance of attracting pollinators. When that happens, chances for successful reproduction are severely diminished.

Hawaii's native insects have long been known to be exotic; some also have been of great use to science. Hawaii is the world's diversity center for the fruit fly, *Drosophila*. Perhaps no other creature on the earth has taught us so much about genetics and evolution as has this small fly. Unfortunately for drosophilids and other insect species known only from Hawaii, ground-nesting yellow jackets were introduced to Hawaii in the 1970s. These bad-tempered wasps are insect predators; they consume huge numbers of local insects each day. Since their introduction, entomologists have noted an alarming drop in the population densities of many local insects, *Drosophila*s included.

The list goes on and on. Aphids, mealy bugs, termites, roaches, even European and American honeybees are all destroying or competing with local insect faunas and the plants they pollinate. Mankind has let loose in Hawaii a great number of plagues, and they resist being stopped or controlled. The devastation is immense, ongoing, and unstoppable. Some number of the local insects will survive this alien onslaught. Some, and the trees and other plants they pollinate, will not. But the biggest tragedy is that some unknown number of the native insects are already extinct and will never be known to us.

4

Of all the creatures affected by the arrival of mankind in the Hawaiian Islands, two groups stand out: the land snails and the birds. The actions of humanity leading to the sad history of the snails almost approaches farce; the story of the birds is only cautionary tragedy.

Mollusks are the second most diverse group of animals on the earth; only the arthropods have evolved more species. Using a most unlikely body plan of mucus-laden flesh dragging about a protective shell, the mollusks have colonized virtually every sea, river, and lake on the earth. But they have not done as well on land, and it was not until relatively late in their history that one group, the snails, evolved lungs and crawled onto dry land.

Most mollusks are relatively inconspicuous creatures and, save for the fast, intelligent octopuses and squid, neither very fast nor very bright. But they all have at least one conspicuous evolutionary trait: Mollusks are capable of rapidly producing many new species when given the chance.

That chance was achieved with the formation of the Hawaiian Islands. Several times in the past, large logs or floating branches must have washed onto an ancient Hawaiian Island shore carrying land-dwelling snails, rafted in from some distant island group or continent. These snails, upon reaching Hawaii, found conditions ideal for growth and evolution. By the time mankind arrived in the Hawaiian Islands, the handful of refugee arrivals had evolved into more than a thousand species of land-living snails.

Land snails are much beloved by scientists studying evolution and evolutionary change, for their rapid rates of species formation and evolution can be studied readily both in modern populations and from the fossil record. Most of us dealing with land snails on a day-to-day basis, however, view them only as garden pests, or perhaps as hors d'oeuvres in a French restaurant, and in many eyes the common species found in North America and Europe are only drab and rather loathsome receptacles of mucus. But tropical environments have a way of taking drab species and adding color and variety. The shells of most Hawaiian land-snail species are richly colored, and many show a pleasing array of ornament.

Like so many of the other species evolving in Hawaii, the land snails benefited greatly from the virtual lack of predators. They were rather tranquil beasts, moving little in a lifetime. Two types existed: those living in the forest floor and those living in trees. Most of the snails consumed microscopic fungi and algae, and the tree dwellers caused no damage whatsoever to their host trees. All in all, it must have been snail heaven.

All that changed with the arrival of the Polynesians in Hawaii, for these first Hawaiians unwittingly brought with them a wily predator: the Polynesian rat. Rapidly spreading from island to island, and breeding like only vermin can, the rats quickly made life miserable for the snails. We have reliable evidence of this invasion, for the oldest archaeological digs yielding human artifacts also have produced fossil snail shells showing characteristic break patterns that form only when snails are attacked and eaten by rats. Probably the rats caused a major extinction among the ground-living snails and lesser damage to the tree-dwelling forms. Currently no one has attempted to determine how many ground-dwelling snails were driven to extinction by the combined effect of early Hawaiian agriculture and the introduction of rats into the forest floor ecosystems.

The native Hawaiians had better things to do than collect and count the number of snail species found on the islands, and thus we have little idea about the diversity of land snails present on the island when they arrived, nearly two millennia ago. Species collecting and counting, however, was a favorite pastime of the European naturalists who eventually arrived in the islands. Studies on the snail species living in the Hawaiian Islands began in the 1800s; to date malacologists have identified 931 species and 332 subspecies of Hawaiian land snails, all living during the late nineteenth century. These figures are astounding. In comparison, the entire land area of the continental United States and Canada currently has a total of 719 species of land snails *combined*. But the number of land-snail species currently living on Hawaii is far less than it was during the nineteenth century; snail diversity has dropped grievously, and continues to do so. There is a very real possibility that the number of types of land snails living in Hawaii will have dropped from a thousand to ten in one century.

Several factors have caused the extinction of the land snails: first the rats, then deforestation and other habitat and vegetation change, then overgrazing by goats, sheep, pigs, and cows. But if that weren't enough, the snails have been victimized most recently by the two groups that supposedly love them the most: shell collectors and government scientists.

Starting about 1850 and ending around the turn of the century, a land-snail collecting craze swept the Hawaiian Islands. Private and scientific collectors alike combed the snail habitats to make ever larger shell collections; some collectors hired native Hawaiians to scale the trees, beat the brush, and climb every mountain necessary to find the snails. In a sobering article by University of Hawaii biologist Michael Hadfield, published in 1986, it was estimated that as many as 2,000 snails could be found in each tree at the end of the last century; collectors would commonly return from trips with several thousand snails each. Many of these snails were for private shell collections, but scientists also participated in the catastrophic collecting. One of the worst collectors was a man named J. T. Gulick, who collected and then killed over 45,000 snails in three years. Gulick, who was also a missionary, even bragged in a letter about his pillages, stating that only the extinction of the snails would ensure that his collection remained unique. One wonders if Brother Gulick's methods for saving lost souls were as catastrophic as his efforts on behalf of lost snails.

Many of these gigantic snail collections have, over the years, found their way into the Bishop Museum in Honolulu. I was able to spend a morning looking at these old collections, lined up in box after box, most useless because they lack even rudimentary locality data. In the company of curator Rob Cowie, I examined some of the hundreds of thousands of shells collected over the years. Many of these collections represent the last known individuals of recently living but now extinct species, and the collections themselves are endangered. They are deteriorating rapidly due to a chemical reaction with the material in which they are stored. Cowie has estimated that at least $500,000 will be required to house the gigantic collection properly. It is a hideous irony; some last members of species found nowhere else on the earth have been driven into the black pit of extinction by shell collectors, and now even the shells of these collections are dissolving away.

Snails are perfectly suited to being wiped out suddenly. They are slow moving, not widely distributed, and produce on the average of one or two baby snails each year. (In comparison, any oyster produces about 5 to 20 million babies each year.) This mode of life and reproductive strategy worked perfectly well in the prehuman days, but has not served the snails well since. Even so, perhaps half of the snail species would have survived but for one of the truly classic blunders of modern science, conducted in the mid-1950s.

During the first part of this century, some fool imported a creature known as the giant African snail into Hawaii. This snail, a gastronomic delicacy, promptly escaped and began devouring local crops with great gusto. People tried to poison these snails, but when lots of poison snail pellets began to kill off more local birds than snails, some genius in Hawaii's Department of Agriculture decided to import a carnivorous snail to combat the ravages of the African snail. Malacologists on the islands and mainland who were consulted about this move let out a unanimous yell of protest. Nevertheless, two carnivorous snail species were duly imported from Florida and released into the countryside. Unfortunately, no one took the trouble to find out if these carnivores really ate giant African snails. Much to the consternation of the Department of Agriculture, it was soon ascertained that the carnivorous snails preferred the native land snails over the target species. The giant African snails were left to their own devices, which was to devour crops with astonishing speed, while the rapidly multiplying carnivores wreaked havoc and extinction among the placid, tree-dwelling algae eaters. The result has been a monstrous and monumental extinction. By 1970 it was estimated that 50 percent of the approximately one thousand snail species known to have been living in the Hawaiian Islands during the first part of this century were extinct, and in some cases the extinction rate was far higher. On Kauai, one particular snail genus was known to be composed of twenty-one species late in the nineteenth century; by 1931 only eleven of these were still living; and by 1970 only a single species could still be found alive. Worse, the extinctions continued to occur at an ever increasing rate. By 1981 the branch of the United States federal government involved in wildlife protection, the Fish and Wildlife Service, finally smelled the coffee and declared an entire genus of land snails to be endangered; twenty-two species assigned to this genus were recognized as newly extinct, and the other nineteen species were recognized to be on the brink of extinction. In 1990 it was estimated that a further 25 percent of the land-snail species had gone extinct since 1970.

Biologists studying these extinctions universally agree that the introduced carnivorous snails have been the primary culprits in snail extinctions since the middle part of this century, and remain the single worst threat to the few remaining native Hawaiian snails. The introduction of these carnivores by biologists belonging to a state agency in Hawaii was a stupid, tragic miscalculation. But the whole affair, in my mind, is elevated into the realm of very sick farce by the response of the biologists who brought about this disaster: They announced to the world that their giant African snail eradication program had worked and successfully talked other island nations in the Pacific region into introducing the same predatory snails.

The most vicious and predatory of the two carnivorous snails introduced into the Hawaiian Islands came from Florida; its scientific name is *Euglandina rosea.* In 1977 the French government purposefully introduced *Euglandina* into the island of Moorea, near Tahiti, also in the hopes of controlling the giant African snail. Once again, the predatory snail neglected its mission in favor of the apparently more delectable native snails. Three American scientists went to Moorea to study the rate at which *Euglandina* disposed of the Moorean snails. Within ten years all of the native Moorean land-snail fauna, composed of a dozen species in 1977, had been completely eradicated from the face of the earth. Never before in history had a more savage, rapid extinction been witnessed and documented directly by scientists. The same predatory snail has been recently released in Tahiti and Guam, among other islands, and has nearly eradicated the land-snail faunas there as well. A great diversity of life has been forever lost from the face of the earth.

5

There is a tremendous bias about which endangered species receive conservation efforts and which are left to their fate. Large, endangered mammals and pretty birds such as spotted owls receive the highest priority, followed by other vertebrates such as reptiles, amphibians, and fish. Invertebrate animals and plants are always far down the list. Some conservation groups are concerned exclusively with high-profile animals; the World Wildlife Fund, for instance, one of the largest agencies in the world trying to rescue endangered species from extinction, does not recognize a single invertebrate species as worthy of help. Yet vertebrates are only a tiny fraction of the earth's biota, and in many cases they are far better off than many of our world's smallest creatures. Saving invertebrate species often devolves onto individual, interested conservationists. In Hawaii, the fate of the land snails is currently in the hands of a single man.

In early February of 1992 I was able to visit with this man, Professor Michael Hadfield of the University of Hawaii. Hadfield and I have much in common; both raised in the Northwest, and both graduates of the University of Washington, we had also benefited from the sage advice and teachings of the same mentor, a great invertebrate zoologist at the University of Washington named Paul Illg. It was from Hadfield that I learned much about the snail wars currently under way in Hawaii and on other islands in the Pacific region. After two decades of field studies, Hadfield has learned enough about the natural history of numerous endangered snail species that he has been able to begin a successful artificial rearing project. But Hadfield is doing more than simply raising snails in the hopes that someday they can be released back into the wild. As long as the carnivorous snail *Euglandina* is alive in Hawaii, the native land snails will remain endangered. Hadfield has therefore taken the fight to the carnivore itself.

For several decades *Euglandina* has been migrating ever higher into the Hawaiian mountains. Today local populations of native snails can be found on trees only at the highest elevations; they are like frantic survivors climbing ever higher to avoid a rising flood. But the carnivorous snails have nearly reached the last, highest outposts. Hadfield has recently discovered a poison that is specific for the carnivorous snails, and this effort is the last hope. If it fails, the once-prolific fauna of Hawaiian snails will be devoured into extinction by the end of the twentieth century. Over a thousand species will have been destroyed from a single, small group of islands.

6

Great devastation was unleashed upon the Hawaiian biota following the arrival of European civilization, and until about 1980 it was assumed that the vast majority of extinctions took place during the last two centuries. In 1982, however, this view forever changed.

The island of Kauai is one of the world's beautiful places, rightly called the Garden Island because of its rich, lush cover of vegetation (most of it now composed of alien, introduced species). Its coastline is a mixture of black volcanic headlands and pristine beaches; one of the most beautiful of these beaches is a place called Poipu Beach, where snorkelers and boogy-boarders now compete for space in the clear blue sea. If you continue eastward from Poipu Beach, the landscape changes somewhat; the low, black basalt created by the island's primeval volcanos is overlain by giant, golden sand dunes, some new, others already solidified into rock. Along this coastline, near a place called Maka'wehi, a beachcomber in the 1970s noticed some thin white bones eroding out of the dunes. This chance discovery resulted in a complete reevaluation of mankind's effect on wildlife. The bones came from large, flightless birds belonging to species no longer existing on the earth. But they were not of species inhabiting the earth millions of years ago. These great birds lived more than 200,000 years ago, and still did so when Christ walked the earth, and later yet, when Rome fell. But they, and many other bird species as well, went extinct soon after the arrival of the first Hawaiians.

7

Museums are among the oddest institutions ever invented. They are simultaneously storehouses of information and places of public learning or pleasure; organizations promoting research as well as vast repositories of public property. Many museums, however, also are becoming the messengers of bad news, a last chance to see rare and vanishing treasures; it is the museums that document how much we have lost and how much we still stand to lose. Many museums are now sad graveyards, holding the bones of creatures both fabulous and ordinary that have died in the current extinction, as well as the stuffed bodies of other creatures soon to join the lists of the dead.

On a warm, late-autumn day I journeyed into the back regions of the Smithsonian Institution to see such victims. My normal haunts in this fabled museum are among the paleontological collections, where giant rooms hold seemingly endless cases packed with fossil treasures, creatures so long dead that they are difficult to conjure as living even in the most active imagination. But on this particular day I took the plodding elevators up to the Department of Ornithology, to visit with a man and see the collections he has helped gather, collections that have tragically revolutionized our understanding of the Third Event. So much is seen in the popular press about the possibility of an impending extinction, and about how many species still will be around by such and such a date. But the work of curator Stors Olson and his wife Helen James on the extinct bird faunas of Hawaii has made such questions moot. Learned commentators pontificating about "when" the extinction will occur should be asking, "How much has already disappeared?"

I had first met Stors Olson in Seattle, on a cold blustery day in the early 1990s, after he had given a seminar about the extinctions of birds on islands during the last 10,000 years. Up until that time my work and thoughts about extinction dwelt far in the past; extinctions were things that happened long ago, or would happen sometime in the future. I knew, as we all do now, about rain forest destruction, and I believed the various arguments positing an impending extinction in the tropics. But I was unprepared for Olson's message; I was under the impression that other than the loss of a few large mammals during the Ice Age, the world had not experienced any significant loss in diversity for millions of years. It was therefore a great shock when Olson calmly told his audience that over *half* of all the birds on virtually every one of the world's many islands have gone extinct relatively recently. About 8,500 species of birds are now known on the earth. The islands of our world hold about one-sixth of this total. If Olson is correct, as many as 1,500 birds have been eliminated from the earth in the last few thousand years. I remember sitting back in my chair, trying to comprehend this number. Fifteen hundred species is far more than all of the known dinosaur species put together from a 120-million-year history. It is about half the number of mammals now living. And if this many bird species have so suddenly and so recently disappeared from the earth, how much else that has not left behind a fossil record has gone extinct as well? It was this message, more than any, that started me on the long path leading to this book.

The work of Olson and James was revolutionary not for its methodology, which was simple enough; they went to Hawaii and patiently excavated fossil bird bones, and then compared the reconstructed species with those still living in Hawaii. The import of their work came from its sheer unexpectedness. When Captain Cook arrived in the Hawaiian Islands in 1778, the bird fauna contained approximately seventy native bird species; this number seemed quite diverse given the relatively small size of the islands and their great distance from other landmasses. In the two centuries since Cook's arrival, sixteen bird species have gone extinct and a further twenty-four are listed as rare or endangered. This tragic extinction was blamed on the habitat destruction due to deforestation and the rise of European agriculture, introduced predators (especially the mongoose, another predator greatly preferring native wildlife over its intended victim, the rat), loss of the birds' native food due to insect extinctions, and the introduction of avian malaria and pox. None of these post-1800 avian extinctions really came as much of a shock to anyone, however; progress has its price, or so the message went. It was therefore a bolt from the blue when Olson and James announced in *Science* magazine that, soon after the arrival of the first Hawaiian people nearly 2,000 years ago, a far greater avian extinction had occurred.

PHOTO BY THE AUTHOR.

Bird fossil site, Kauai.

Olson and James have now identified over fifty species of birds that were victims of this early extinction, and the number keeps climbing with each year's field work.

Half of Hawaii's native birds disappeared soon after the arrival of humans in the islands. In retrospect, it is not difficult to see why. But in 1982, when Olson and James first reported their findings, it created a bombshell in the communities of native Hawaiians, who had long been proud of their ancestors' ecological practices. Now it appears that a great slaughter occurred.

Olson and James arrived at this conclusion by painstakingly collecting from sites straddling the arrival of mankind. Lava tubes, karst sink holes, and ancient sand dunes have all yielded the delicate, hollow bones of fossil birds. The oldest of these deposits, about 200,000 years old, yielded essentially the same avifauna as sites deposited about 2,000 years ago, immediately prior to humans' first arrival in the islands. Sites deposited soon after their arrival show a vastly different and far smaller assemblage.

The entire Hawaiian native bird fauna apparently evolved from about twenty colonizing species. These chance colonizers, which arrived from far lands, included a heron, an ibis, several different geese and duck species, at least three kinds of rails, several predatory birds including hawks, owls, and an eagle, and several types of forest birds. Most of these birds rapidly diversified into numerous new species. And because the Hawaiian Islands had no predatory mammals or reptiles, many of these new bird species became flightless.

The flightless birds of Hawaii would have been extraordinary to see. I was able to look at their bony remains in the Smithsonian, and was surprised at the large size of many of these extinct birds. At least twenty-five flightless species evolved from the various water fowl, producing a variety of giant, waddling ducks and geese. A large ibis, also flightless, may have been the largest bird on the island. When the first Hawaiians arrived they found flocks of striding, walking, waddling, preening, and ultimately helpless, flightless birds, all soon running for their lives in the face of hungry humans and dogs. The flightless birds had no chance and were probably gone very soon, for the islands contained very little edible vegetation or other protein; to the early Hawaiians, the slow, stupid birds must have seemed like manna from heaven. Also gone very quickly were several species of hawks, owls, and eagles, birds that apparently could not tolerate habitat destruction and the sudden loss of their usual prey.

We will surely never have more than an approximate idea of the Hawaiian bird fauna's diversity prior to the introduction of mankind; surely many rare species left few or no fossils at all. After visiting Olson's various collecting sites around the islands, I am amazed that he and Helen James found as many species as they have. Birds rarely fossilize; since their bones are light and hollow, they are usually destroyed by burial, compaction, and dissolution. That so many fossil bird species have been recovered to date is a testament to the dogged, detailed field collection of Olson and James. But it is also a statement about how rich the bird fauna of Hawaii once was. We know, at best, that many extraordinary birds lived on Hawaii very recently. We know that they were species found nowhere else on the earth. We will never know, however, their colors, or calls, or the dances they used to win mates; we will never see the form of their nests, or the food they ate, or the color and shape of their eggs. The traits that made these birds living creatures, the things we would want to see of them, are lost forever. Extinction leaves only bleached broken bones in the best of cases.

Flightless birds were not the only ones to succumb. Many forest birds rapidly went extinct as well, apparently from the effects of habitat destruction. Hawaiians also hunted many of the brightly colored forest birds for their feathers. I was appalled to learn that each of the feather capes worn by the Hawaiian monarchs required the feathers of 80,000 birds. These capes, one of the proudest cultural adornments of the Hawaiian people, surely pushed more than one species to—or over—the brink of extinction. Today, on virtually every large Hawaiian street of every island, the libraries, schools, and public attractions are all marked by a sign bearing an idealized Hawaiian king wearing a cape of pink feathers. Whenever I saw these signs I was reminded of what there once was and what there is now in the way of birds on Hawaii. The only birds I saw during my stay were alien species, such as sparrows, doves, and mynahs. Of the original native bird fauna, surely numbering more than one hundred species before humans arrived, only nine now exist in populations large enough to suggest that they will survive. The rest are extinct or nearly so. No one is to blame. Everyone is to blame.

The work of Olson and James encouraged other paleontologists to examine other islands around the world, and the same pattern soon emerged. In every case where fossils could be found, large numbers of extinct species were soon uncovered. Where reliable dating could be ascertained, it was found that the extinctions always occurred soon after the arrival of mankind and that about half of the native faunas were always lost. Those species still left today are the hardy forms and the weeds.

Just before taking leave of Stors Olson in his museum refuge, I asked him one final question. "How did the Hawaiians take the news?" I asked. Olson shrugged. "I'm not the most popular man in Hawaii right now," he replied.

8

Probably very few of us do not indulge in island dreaming, especially on dark, winter days. For many years my dreams centered on Madagascar, a giant island inhabited by some of the world's most unusual creatures, including our prosimian primate ancestors, the lemurs. I envisioned it as an exotic, tropical land, lying as it does off the southeast coast of Africa; I was sure it would have the requisite golden beaches and warm sunshine, the two antidotes necessary to vanquish the winter blahs. But these attributes would have been but welcome bonus attractions, for my primary interest in Madagascar lies in its fossils, not its living treasures: Madagascar has more Mesozoic ammonite fossils than any other locality on the earth. Much of this rich fossil treasure had been accumulated by the mid-twentieth-century efforts of one man, a French army general named Maurice Collignon, who used his troops to collect fossils instead of suppressing revolutions. Some aspect of ancient Madagascar must have proven especially beneficial to my lovely, long-dead chambered cephalopods, for they occur there in prodigious numbers. During the travels of my life, however, I never seemed to get near this fossil-rich island. But such paleontological largesse was too strong a temptation to deny forever, and I was finally poised to arrive in this promised land in the late summer of 1991. My arms covered with shots, my blood filled with protection against yellow fever, tetanus, cholera, and my system positively buoyant with gamma globulin and antimalarial medicine, I waited in South Africa, ready for the short hop over the Mozambique

Channel. At the last minute, however, a bloody revolution toppled the corrupt Malgache government, and the death toll from those gunned down in the streets during the insurrection (as well as dire warnings from the U.S. State Department) dissuaded me from pursuing my paleontological studies in Madagascar. Frustrated from making an actual visit, I had to content myself with looking at the island from afar. In retrospect, perhaps, it is as well that my trip was aborted; from the vantage point of my university library, I discovered my island dream to be part of a much sadder reality, with a story of deforestation and extinction frightfully similar to but even more devastating than Hawaii's.

Madagascar is the fourth largest island on the earth, with a length of over 1,000 miles and a land area about the size of Texas. It has long been separated from the African continent, and because of this, the evolutionary history of the Madagascar biota is very different from that of its nearby continental neighbor. Madagascar contains nearly 10,000 species of flowering plants, of which four-fifths are endemic; half of the world's chameleons and most of the world's prosimian primates are found there as well. Unlike Hawaii, Madagascar contained numerous native mammals and became home to a rich assemblage of unique mammalian species. But those animals and plants still alive on Madagascar today are but a fading echo of the bestiary that existed there only 1,000 years ago, for Madagascar was discovered by humans about 1,500 years ago and, soon after, the extinctions of its unique fauna began with tragic rapidity.

As in Hawaii, the discovery of Madagascar's past faunas came from fossil digging. In Madagascar, however, this study is nowhere near as advanced as in Hawaii; we are only just beginning to see what was present prior to mankind's arrival and what has been lost since. Most of the fossils come from only a few sites, and the record gleaned to date is strongly biased toward larger animals, apparently the forms most likely to have fossilized. As yet, there is little or no knowledge about the disappearance of smaller species.

Archaeological discoveries have ascertained that humanity arrived late to Madagascar relative to most other parts of the world; no humans appear to have settled there permanently until about A.D. 500. By about A.D. 1000, the wave of extinctions following mankind's arrival in Madagascar was complete. The list of victims, even at our preliminary state of knowledge, is long.

Perhaps the most tragic of the extinctions occurred among our ancestors, the lemurs. These primates are quite distinct from the more familiar monkeys and apes; the narrower faces and long, furry tails of lemurs set them apart completely. The still-living species probably look much like fossil lemurs recovered from Early Cenozoic strata, and thus can be considered as living fossils. Although lemurs once had a worldwide distribution, today they are restricted to Madagascar and a few neighboring islands.

A minimum of seventeen lemur genera and many times that number of species were present on Madagascar prior to mankind's arrival. This figure is a world's record, for no other place on the earth, in the past or present, has contained so many sympatric primates. Of the original seventeen genera, only ten still survive, and two of these have lost at least one species. All of the extinct forms had one trait in common: All were large, with the largest of the extinct forms having a maximum weight of 200 pounds. In contrast, the largest living lemur found today checks in at under twenty-five pounds, and most other species are far smaller and lighter than this. Those that survive are all nocturnal and arboreal; if any terrestrial species once existed, it has been eradicated completely. Because of the poor quality of the fossil record, we have no idea how many small species may have gone extinct as well.

The elephant birds were the other major group to have disappeared. Between six and twelve species of these large, flightless birds lived on Madagascar as recently as a thousand years ago. They must have been among the world's most extraordinary birds. The largest species was about ten feet tall, and massive. (For comparison, *Sesame Street*'s Big Bird is a little smaller than an adult elephant bird.) Elephant birds were important parts of the herbivorous community, and were grazers or forest browsers. They must have produced extraordinary eggs as well, for numerous eggshells are still found on many Madagascar beaches. The eggs, when unbroken, held about three gallons of liquid within. (I can visualize it on a menu: deluxe one-egg omelette for ten. Apparently the ancient Malgache people could visualize this great meal as well, which is precisely why there are no more living elephant birds.)

Other extraordinary creatures were lost too. A pygmy hippo, a giant tortoise, an endemic aardvark, and a large, catlike viverrid carnivore all disappeared within five centuries of humanity's arrival on Madagascar. How much else disappeared is as yet unknown. There is no record of how many small birds, insects, land snails, frogs, lizards, snakes, micromammals, bats, and plants also have been obliterated from Madagascar. Helen James has recently begun collecting in Madagascar, however, and we soon may learn the bad news about smaller bird extinctions. For the other groups, there may never be a gravestone. The large species in any ecosystem are like the canaries carried by ancient miners: Their deaths indicate a poisoned environment. We have documentation of the extinction of about one hundred species on Madagascar since people arrived there. But how many really became extinct? A thousand species? Five thousand? More? When an area the size of Texas, once home to a rich and endemic biota, is burned to the ground, how many species found there and only there actually become extinct? We cannot know. We will never know.

As in the case of the continental Overkill controversy, there is no end of debate about the cause of these extinctions. We know that the early colonists to Madagascar were not bloodthirsty hunters but part of a pastoral, agricultural society. For the first half-millennium of human occupation on the island, population numbers remained low. Hence it has been argued that humans could not have caused the rapid extinctions that swiftly occurred. It has been theorized in some quarters that the extinctions affecting Madagascar were due solely to climate change. On the other hand, there is good evidence that the arrival of mankind on Madagascar set off a wave of vegetation change, including the deforestation of significant proportions of the island, brought about by slash-and-burn agriculture as well as by overgrazing by introduced cattle. Most scientists dealing with the issue suspect that some combination of habitat change due to agriculture coupled with adventitious hunting by the human population is to blame. Robert Dewar, a scientist who has spent many years studying the Madagascar extinction, has proposed a scenario that goes beyond a simple "People arrive, things die" viewpoint. Dewar has pointed out that prior to humanity's arrival, there existed a unique herbivore community composed of elephant birds, tortoises, and lemurs (among others). All of these species were large, lived at relatively low population numbers, had low reproductive rates, and were slow and easy to hunt. The earliest humans on the island were cattle pastoralists, who traveled with their herds across the island. They killed game on occasion, but much of the damage was done by their herds, for cattle can disrupt completely the pattern of native vegetation. Rapid changes in vegetation spelled doom for the native herbivores. The lesson is all too clear: Humans can drive species into extinction by their very presence. It does not take concerted hunting or even very many humans to wipe out wild species. In the case of Madagascar, the ecosystem was disrupted at two levels: disturbance of the native vegetation and removal of native consumers in the trophic pyramid. As in any house of cards, knocking out the base brings down the upper levels.

The human population of Madagascar remained relatively low into this century; around 1900, only an estimated 2.5 million people lived there. But as in many other places on the earth, the population began to rise rapidly during this century. By 1950 the population had doubled, and by 1987 it had doubled again. By the year 2020, twenty-six million people will live on Madagascar. The island is already considered one of the most devastated habitats on the earth, with some scientists estimating a 90 percent deforestation since the arrival of people. It is also one of the poorest places: The per capita income is about $250 per year and declining. As the human population grows, so too will the floral and faunal extinctions.

Madagascar is not the only giant island to have undergone devastating extinctions soon after the arrival of mankind. New Zealand, only slightly smaller than Madagascar, also has lost some extraordinary creatures, and surely many more meek and tiny ones as well. The most spectacular of these is a giant bird known as the moa.

The first fossil remains of moas were discovered in New Zealand in 1830, and since that time there has been unending controversy about what killed them. Once again, the discussion has centered on whether they were killed off by people or if they were simply outmoded animals due for extinction anyway. (One biologist named Roger Duff characterized them as birds developing the fatal New Zealand tendency to adopt a pedestrian habit.) There were many species of moas on New Zealand —between thirteen and twenty-seven have been identified, depending on which taxonomist you believe—and some of them were huge: The largest may have been ten feet tall and weighed about 500 pounds. They were peaceful herbivores, in all probability, and evolved in an ecosystem without large mammalian carnivores.

Although they were all gone when the Europeans first reached New

Zealand, we just missed the moas, for the last individuals appear to date back as recently as A.D. 1500 to 1800. Moas were quite plentiful, however, when the first Polynesians arrived in New Zealand about a thousand years ago. There is no doubt about what killed off the moas, for numerous archaeological kill sites filled with moa bones have been found. Early in their occupation of New Zealand, the Polynesians also burned much of the vast islands; virtually the entire eastern side of the South Island was burned to create cropland, as were the eastern, central, and northern regions of the North Island. Moas were not the only birds to fall victim: The total number of bird species now known to have gone extinct in New Zealand since mankind's arrival there now stands at thirty-four, and this number is surely a minimum estimate; as on most other islands, the study of avian paleontology in New Zealand is still in its infancy. We have no information on what else has disappeared besides birds.

The list of island extinctions is as numerous as the list of islands on our earth. On the Chatham Islands near New Zealand, between nineteen and twenty-seven species of birds have gone extinct. On Easter Island, the Polynesians eliminated virtually the entire flora, then eventually starved to death after eating most living things off the face of that bleak isle. In the Indian Ocean, Europeans killed off the dodo, while in the Atlantic, the great auk was exterminated in the arctic and rails disappeared from Ascension Island. What other creatures were victims is anybody's guess. Most islands have not been studied. Undoubtedly much bad news awaits as paleontologists expand their search.

9

The snail and bird extinctions documented from Hawaii and other islands are known for a single reason: The extinct groups left fossils behind. Without the ancient snail shells left in the Hawaiian rocks and in old collections, and without the delicate bird bones found in scattered locations around the islands, we would have had no idea of the great diversity of these creatures prior to the arrival of mankind. But these two groups are about the only creatures of ancient Hawaii that *have* left fossils. Are we to assume that they and they alone suffered extinction immediately after people arrived? Can we assume that the myriad species of insects, spiders, crustaceans, plants, and other small creatures without skeletons existing here two millennia ago survived without loss? Such a suggestion is ludicrous. A gigantic extinction has taken place on Hawaii, on other islands, and on other continents. There are no villains, except, perhaps, for the people who foolishly introduced creatures to feed their vanity or pocketbook. People arrived, and species died. Hawaii offers a tremendous lesson; it shows that many species on the earth cannot tolerate the least human disturbance, so delicately are they balanced on the precarious tightrope of nature. Hawaii tells us that the Third Event is not only something to fear in the future; it has been long under way.

The good news is that many men and women living today in the Hawaiian Islands are trying to save and preserve the remaining native species, treasures far greater than all of the golden beaches and jungle waterfalls of this island paradise. The bad news is that today Hawaii has the highest number of officially recognized rare or endangered species of any American state. Worse news is that Hawaii is one of the few islands on the earth that recognizes that it *has* rare or endangered species.

Chapter Eleven

Numbers

I

In a cold midafternoon rain I drove through Seattle toward the airport. As usual, the freeway was choked with traffic; the concept of "rush hour" no longer applies here, for the ever-expanding population in a so-called livable city has created a never-ending gridlock, a "rush day," if you will. In the airport I caught a quick news telecast, noting with relief that we were still at peace. The flight to Chicago was a chance to relax, an oasis of calm between the rigors of departure and the stress of the upcoming few days. My plane arrived in falling snow, and it was immediately apparent that the state of peace had changed, for armed men now guarded the doors of the airport and paced about nervously, not quite knowing what to do. Sometime during the three hours of my flight the United States had gone to war.

My taxi ride into town was set against the backdrop of a blaring radio broadcast. On this cold winter day, the United States had bombed Iraq. Success or failure, further war or quick peace, great national pride or woeful humiliation—ultimately, life and death—was being reduced by the radio commentators to endless varieties of numbers.

I had been invited to Chicago to give two lectures about an older time of death, the 65-million-year-old Second Event, but my motives in coming to this wintry city had as much to do with learning as teaching. The University of Chicago is a distinguished center of paleontological study. Here, better than anywhere else, I hoped to find answers to a troubling question, one that can be answered only through numbers: How catastrophic will the current extinction be, compared to its predecessors?

There are numerous students of evolution and paleontology in the Chicago area. At the University of Chicago, three of them—David Raup, Jack Sepkoski, and David Jablonski—have made fundamental discoveries in the nature of extinction processes. But their work has embraced far more than the history of death. They and other scientists have shown that the current extinction is progressing in a biota fundamentally different—not only in terms of composition, but in more subtle, taxonomic ways—from those toppled by previous extinction events.

As anyone who has read this book can surely now tell, I am an unabashed admirer of my field, paleontology. Relegated to the fringes of the scientific establishment for the first half of this century, a great revolution in paleontological research beginning in the 1960s has thrust the field back to the forefront. One of the major discoveries of this period dealt with the number of species on the earth: As I have documented earlier, the history of life on this planet during the last 600 million years has been one of almost steady diversification, punctuated only by temporary setbacks imposed by mass extinctions. Prior to this discovery, it had long been thought that species diversity had early on reached some maximum level (thought to be imposed by some sort of evolutionary carrying capacity of the planet) and then remained constant. Work by people such as James Valentine and Jack Sepkoski showed this not to be true; we now believe that the number of species in the present day is far higher than at any time during the Paleozoic or Mesozoic eras. But a totally unexpected offshoot of this research was the finding that diversification has been taking place at different rates among the various taxonomic categories: Although millions of new species have been produced over the past 100 million years, proportionately fewer *higher* taxonomic categories have evolved, such as genera, families, orders, and classes. Evolution and diversification have occurred through the creation of new species among already existing body plans, rather than by the invention of entirely new groups.

Species are composed of individual organisms that are capable of interbreeding successfully. Because the evolutionary process causes one or more species to arise from other, coexisting or preceding species, many species share common ancestors. Thus blocks of species are united by common heritage, just as the siblings of a family share a common set of parents. These groups of related species are called higher taxa.

The enrichment of species noted over the two decades by paleontologists, among the higher taxonomic units, was unexpected, and it has produced a bias in estimating the severity of modern extinctions compared to those of the past. The severity of an extinction is a function both of its extinction rate—the number of species becoming extinct per time unit—and of its percent extinction—the number of species suffering extinction divided by the total number of species on the earth at that time. But because most studies compare the losses of higher taxonomic categories, such as genera or families, rather than species, they have consistently led to an *underestimation* of the current rates of extinction compared to the great events of the past.

An analogy can illustrate this process. Let us imagine that each car model currently driving around today is a species, and each company it came from a genus. All belong to one family, the Family Cars. Other families are on the roads as well: the Family Trucks, the Family Motorcycles, the Family Roadgraders, the Family Ambulances, and so on. All of these families first evolved around the turn of this century, and all can be placed into even a higher category, the Order Combustion-engine Vehicle. Since the time of the origin of the Family Cars, the number of species has proliferated enormously; where the genus Ford once had only two species, the Model T and the Model A, it now has the Taurus, Probe, Escort, Thunderbird, Tempo, Mustang, and so on, as well as many extinct species: the Galaxy, Fairlane, Pinto, Edsel, and so on. The result of about ninety years of evolution among the cars is that each car company now offers far more models than it did in 1900; Cars have thus diversified. But the number of car *genera* has increased only slightly in the same period (we now have Hondas, Toyotas, Nissans to go along with the Fords, GMs, and Chryslers), and the number of families has barely increased at all. (The passenger van is one of the only new additions in two decades.) Early in the century, all of the major families, which correspond to distinct body plans—the cars, trucks, motorcycles —soon appeared, and have remained relatively stable in overall design ever since. This is not to say that evolution has not occurred, for cars have evolved enormously in details such as styling and engine type.

Nevertheless, all of the car species still have four wheels, carry passengers in a cabin, and so on.

Let us now compare the diversity of vehicles on the road in 1920 with that of the present day. If we count only families, the numbers would be quite similar. If we count species, the numbers would be very different. Compared to 1920, the number of families may have increased by three or four, the number of genera has increased by several tens, while species numbers have increased by many hundreds.

The diversity of creatures has acted in similar fashion. Compared to the Paleozoic Era, the number of currently living families has increased slightly, but the number of species has increased enormously. And yet most measures of diversity of living creatures through time have depended not on counts of species but of families. In a similar fashion, most estimates of extinction levels during the various mass extinctions also have depended on rates of family extinction, not species extinction. Because of this, the current extinction looks far less severe than either the First or Second Event. At the species level, however, just the opposite may actually be true.

Each phylum (a basic building plan for animals and plants) is made up of one or more classes, themselves divided into orders, then families, genera, and species. During the Paleozoic Era, each genus might be composed of but a handful of species, and each family but a few genera; because of this, during periods of increased extinction, the loss of even a moderate number of species could mean the loss of many families or other higher units as well. As time progressed, however, families have become increasingly and disproportionately packed with new genera and species, each a slightly new variant on an already established body design. Taxonomic groups such as genera and families are now composed of far more species than at any time in the past. These larger taxonomic groups are thus more extinction-resistant than during past eras, since today the extinction of hundreds or even thousands of species may be needed to eliminate a given family. This form of bookkeeping—counting only the families going extinct—while useful in keeping track of the world's creatures, masks the true calamity of the modern extinction.

For instance, in the great extinction ending the Paleozoic Era, the First Event, as many as 50 percent of all *families* of marine organisms became extinct. By comparison, the devastating extinction ending the

Mesozoic Era, the Second Event, was accomplished with only 15 percent of marine families dying off. But because the average number of species per family increased from the Paleozoic to the Mesozoic, the actual *number* of species disappearing in each of these events may have been similar.

Two points are undisputable: The number of species has increased through time, and there are more species per family now than at any previous time. But most people attempting to grapple with the problem of current and impending extinctions have missed these two salient points. All too often they argue that the current extinction is far less calamitous than either the end-Paleozoic or end-Mesozoic events (and thus not worth getting too upset about) because, supposedly, a lower *percentage* of families and genera are now going extinct than in the past. Second, the severity of a given extinction is commonly tabulated as a percentage of extinct taxa compared to the total number of taxonomic units, whether they are families, genera, or species. Using this measure, scientists have argued that the extinctions occurring to date since the onset of the Ice Age have been trivial compared to the earlier great extinctions, because the percentage of taxa becoming extinct is but a tiny fraction of the total diversity of the earth. What these scientists overlook, however, is the fact that the absolute—not relative—number of species (or other categories) that have *already* gone extinct in the last million years may be more than the total of the First and Second events combined. I have tried to tabulate the number of species known to have been exterminated 65 million years ago by the asteroid. The numbers are surprisingly low. We know of perhaps 200 vertebrates, about 100 planktonic foraminifera, around 500 to 1000 mollusks, perhaps an equal number of plants, and several hundred species of many miscellaneous groups—altogether, we can document the death of perhaps 2,000 to 4,000 species in the Second Event. This is not to say that thousands or perhaps even millions more species did not go extinct simultaneously, for surely our fossil record is highly incomplete. But by the same measure, if Stors Olson and Helen James are correct, at least 1,000 species of birds alone have disappeared from the earth in the last two to five millennia, and perhaps an equal number of snails has disappeared just from the Hawaiian Islands in the last millennium. In these two groups alone, we have record of an extinction approaching in severity that of the Second Event. And these are species that have left a fossil record, ones we know about. How many more have gone extinct without leaving a trace? The answer must be many.

Chicago paleontologist David Raup has produced another number relevant to understanding current extinctions: He has introduced a term called background extinction rates: the number of extinctions taking place during "normal" times, when the earth is not suffering a mass extinction. Extinction is the ultimate fate of every species; just as an individual is born, lives out a time on the earth, and then dies, so too does a species come into existence through a speciation process, exist for a given number of years (usually counted in the millions), and then eventually become extinct. Like the obituary page of any newspaper, the fossil record has tabulated random extinctions taking place throughout time. But Raup and others have shown that the rate at which these "random" extinctions have taken place through geologic time is remarkably low. According to Raup's calculations, the background extinction rate during the last 500 million years has been about one species going extinct every five years. But he gets these figures by taking into account *all* extinctions during the last 500 million years, including those during the periods of mass extinction. If we delete the number of extinctions occurring during the mass extinctions from the calculations, the background extinction rate drops to nearly zero. Apparently, during most of earth history, hundreds or even thousands of years may have passed between species extinctions. In contrast, Norman Myers, one of the earliest scientists to warn of a current mass extinction, has estimated that four species *per day* have been going extinct in Brazil alone for the past thirty-five years. Stanford biologist Paul Ehrlich has suggested that, by the end of the century, the extinction rate may be measured in species *per hour.*

These were the issues discussed among my Chicago colleagues. Against a backdrop of Scud missile attacks on Israel, endless television features on oxymoronic "smart bombs" and "military intelligence," and the specter of imminent ground warfare, I tried to keep my mind on the past and on the future. But two events seemed to overshadow the high-tech bombardments, activities that were capable of producing shock even after the travails of the twentieth century: The release of a giant oil slick into the fragile Persian Gulf and the Iraqi firing of the Kuwaiti oil fields, filling the sky with black smoke, seemed sins beyond comprehension. I can somehow understand people killing people; I suspect it has been going on since our australopithecine beginnings. But the wanton attack on the earth with oil slicks and fires provided a stark symbol of what we are capable of doing against species other than ourselves.

2

No one disputes that the activities of mankind have caused extinctions in the recent and not so recent past; the phrase "dead as a dodo" is not pure whimsy. But there is currently a great debate about the extent of man-made extinctions, and even more about the promise—or threat—for the future. Ultimately, the entire issue devolves on numbers. But the numbers we need are very difficult to obtain: How many species exist on the earth? How many have there been at various times in the past? How many species have gone extinct in the last millennium, the last century, or even the last decade or year? And most important of all, how many will be gone in the next century, or millennium, or million years? None of these numbers is directly obtainable; each has to be reached, if at all, by abstraction, inference, deduction, or just plain guesswork. It is no wonder that critics of those trying to tell the world that we have entered a period of mass extinction—and that it will only get much worse unless something is done—are having a field day.

"Species Loss: Crisis or False Alarm?" read the large headline in the August 20, 1991, edition of *The New York Times*. The article was in response to an earlier edition of *Science* magazine, which had devoted a handful of articles to the question of global biodiversity and its potential, impending loss. The major point of the *Times* article was that great skepticism concerning the more catastrophic estimates of species loss seemed prudent, in light of the current very poor understanding of global biodiversity. Furthermore, various skeptics wondered if even 10 percent of current world species diversity would be lost, and suggested that such a small loss would hardly be noticed. This is the very point that remains so poorly understood or is willfully overlooked by those unconcerned about biodiversity loss: A 10 percent loss of species at today's diversity makes the current extinction every bit as severe in actual number of species deaths as any extinction in the past and may, in fact, indicate that a greater number of species is becoming extinct than the combined total for the First and Second events.

Why is there any controversy at all about how many species are currently on the earth? In this day and age, when modern science can detect planets among stars light-years away and can deduce the age and birth of the universe from the movement and activity of subatomic particles, what could be more simple than counting up the number of species, and then, over twenty years, for instance, observing how many become extinct? The answer is, of course, that such an endeavor would require a large army of biologists, instead of the small handful actually engaged in this type of research. In reality, we have only the haziest idea about how many species currently exist on the earth, how many there have been in the past, and how many are going extinct at any given time. This lack of the most basic and vital information—the current number of species living on the earth—is the cause of the great dissension.

Naturalists and scientists have been naming species for about two hundred years, since the great Swedish naturalist Carolus Linnaeus initiated the use of binomial nomenclature and, in the process, began the modern methodology of systematics and taxonomy. Linnaeus's scientific heirs have succeeded in naming about 1.4 million organisms. Of the earth's currently defined 1.4 million creatures, 750,000 are insects, 250,000 are plants, 123,000 are arthropods exclusive of insects, 50,000 are mollusks, and 41,000 are vertebrates. The remainder is composed of various invertebrate animals, bacteria, protists, fungi, and viruses.

Describing and naming a species is a lot of work, and each of the 1.4 million species described to date required just that. You want to be absolutely sure that you are not confusing your new species with one already named, and you have to be sure that all future interested parties will have enough information from your description so that your new species can be recognized. The net result is that only a few new species are described every year. Another problem is that many scientists not involved in biological or paleontological research tend to look down their long noses at those who study and name new species, for taxonomy, the science of naming and ordering the earth's biota, is considered old-fashioned and quite unglamorous. Fewer scientists are capable of conducting this exacting work each year, for very few young people take up the study of classification. In paleontology, very few students devote their theses to the study of species (old or new). Nevertheless, the work progresses slowly.

Until about a decade ago, if you would have asked the average biologist interested in classification "What percentage of the earth's biota has been classified?" you probably would have received some answer ranging from about a half to a fifth; most estimates of the total biodiversity ranged between about 3 and 5 million species. It was therefore a profound shock to systematists the world over to read the rather innocuous report by Smithsonian biologist Dr. Terry Erwin, published in 1982, concerning the number of insect species found in individual trees within the Amazonian rain forest. Erwin suggested that instead of the current estimate of 5 million species or less, there may be *30 million* species of insects alone.

Erwin arrived at this estimate in the following way. He studies tropical insects (mainly beetles), but had been frustrated in trying to collect those found within the higher canopies of rain forest trees in the Amazon Basin. Many of these trees are so tall and their upper canopy so complex that the normal method of collection (you know—hiking around with a little glass pickle jar, ready to yell "Gotcha!" when you pop the lid over an unsuspecting bug) was clearly not going to do. Erwin attempted a new tactic: He decided to increase the size of his jar. Entire trees were covered with plastic and the interior of the giant container filled with bug spray. When the gas cleared, scientists picked up the dead bugs that had fallen out of the trees. (This particular scene was almost duplicated in the movie *Arachnophobia.*) To Erwin's astonishment, the resulting insect collection was at least *ten times* more diverse than he had anticipated. A second shock was his discovery that most of the newly collected insects were undescribed and unnamed—they were species new to science. This methodology was tried on many other trees, in other regions of the tropical rain forests, with similar results: Scientists studying insects, already the single most diverse group of organisms known to exist on the earth, had clearly underestimated species diversity by at least an order of magnitude. Erwin found that many of the new insect species were highly specific to individual trees and that many more of the species had very narrow geographic ranges. It appeared, based on these observations, that evolutionary processes had produced a large number of narrowly adapted and distributed tropical species. The implication for insect extinction was immediately apparent. With so many species found in quite small geographic areas, *any* logging or deforestation is bound to lead to the rapid extinction of many species—most still unrecognized by science.

Erwin arrived at his 30 million insect estimate by extrapolation, for it would take many lifetimes actually to define and count that many different insect species. Yet his estimate, and the effect it had on biologists, cannot be overestimated. It would be like astronomers finding out that the universe was not 15 to 20 billion years old but 150 to 200 billion years in age; or equivalent to physicists discovering that there were ten times as many subatomic particles as previously thought. The word bombshell comes to mind in describing Erwin's discovery, and even a decade later the implications for the extinction debate are just now being realized.

Could there always have been huge numbers of unrecognized insects, even deep in the past? Insects fossilize only rarely, and the diversity of species may never have been estimated correctly. This possibility seems unlikely to me, however. Insects evolved relatively late in the history of life, with most groups appearing either late in the Paleozoic Era or early in the Mesozoic. But by far the greatest stimulation to insect evolution may have been the evolution of flowering plants in the Cretaceous Period and later. Millions of species of insects today are involved in pollination of flowers. Flowers did not exist on the earth until 100 million years ago, so it seems likely that much insect evolution has occurred in the last 100 million years.

If attaining a reliable estimate of global species diversity has caused problems, estimating current rates of extinction has been no less controversial. Much of the problem stems from the fact that so much is at stake—socially, economically, politically as well as biologically—that the numbers and estimates carry enormous political ramifications. Those attempting to preserve species (even if this means slowing the rate of economic growth) use one set of numbers; those seeking to continue economic growth, especially in Third World, tropical countries, counter with a different set. A book recently published by the United States National Academy Press and edited by noted Harvard entomologist Edward O. Wilson assembles a large number of these estimates. One of the volume's authors, Ariel Lugo of the U.S. Department of Agriculture, has charted many of these estimates. Lugo counters the more negative estimates by pointing out the example of Puerto Rico. There, bird extinctions, in spite of deforestation, have not progressed at the catastrophic rates envisioned by many conservationists. On the other hand, although much forest destruction in Puerto Rico has occurred in this century, much also occurred during the earlier five centuries of human occupation of the island. Unlike Hawaii, where the fossil record has shown how many birds went extinct soon after humans arrived, there is as yet no documented avian fossil record from Puerto Rico; we thus have no idea how many bird species were there originally.

If many people disagree strongly on the number of species on the earth and on the rate at which these species are currently declining in number, on one issue there is no disagreement: The vast majority of species currently living on the earth are found in the tropics, mainly in rain forests. The earth's great tropical forests, last legacy of the Mesozoic world, are themselves the most endangered entities of all.

3

Tropical rain forests are characterized by a high canopy, often 30 to 40 meters above the ground with emergent trees towering to 50 meters, and then two or three separate understories of vegetation. They are complex, layered communities with enormously different and changing environments and microclimates. Gordon Orians of the University of Washington has estimated that there are six to seven acres (or 2.5 to 3 hectares) of leaves for each acre of forest.

The term rain forest or, more accurately, tropical rain forest comes from a German botanist named A. Schimper, one of the great naturalists of the nineteenth century. Schimper realized that there are far more than rain forests in the tropics—he identified monsoon forests, savannah forests (also known as dry forests), and thorn forests as well. But when people think of the tropics, they usually think of rain forests, the most species-rich and diverse habitats on the face of the earth. The combined area of these tropical forests covers about 7 percent of the earth's land surface. A minimum estimate is that they contain half of all the world's animal and plant species.

Tropical rain forests require very special conditions: They need constant heat and, of course, moisture—lots of moisture. Rain forests develop where every month is wet, with a minimum of about 25 inches of rain *each month*. Where there are long, seasonal dry periods, true rain forests cannot develop.

Tropical rain forests are today found in three principal regions. The most extensive is the American or neotropical rain forest region (centered in the Amazon Basin but extending up the Caribbean slope of Central America to southern Mexico), comprising about half the global areal total and about one-sixth of the area of all broadleaf forests in the world. The second large block occurs in the eastern tropics and is centered in the Malay Archipelago. The third and smallest block is in Africa, centered in the Zaire Basin.

There is no doubt that the area of the world's tropical forests is rapidly decreasing through human-induced deforestation. There is also no doubt that deforestation leads to extinction—if nothing else, the lesson from Hawaii shows that even slight disturbance to native ecosystems through field formations and forest burning leads to the extinction of the more sensitive species. What *is* currently in doubt is the rate of deforestation and how that translates into a rate of species extinction. Norman Myers, one of the first people to call attention to the link between tropical forest destruction and extinction, estimates that between 45,600 and 55,200 square miles of tropical forest are lost each year through logging and field clearing and that an additional 60,000 square miles are grossly disrupted. This equates to 1 percent of the world's tropical forests disappearing each year; if current practices continue, all tropical forests will disappear in one century. Biologist Edward O. Wilson, in his recent book *The Diversity of Life,* has estimated the rate of loss in 1989 to be a staggering 1.8 percent per year. At this rate, by the end of the current century, only two large blocks may be left: one in the Zaire Basin, the other in the western half of the Amazon Basin in Brazil and Peru. The remainder of the rain forest will exist in fragmented patches. The Food and Agricultural Organization (FAO) of the United Nations officially places the deforestation rate as 0.5 percent per year. Ariel Lugo disputes both these estimates of deforestation and other, even more extreme estimates proposed by Paul Ehrlich. Lugo, one of the most conservative scientists in the extinction debate, suggests a somewhat lower rate. Nevertheless, even with the most conservative of estimates, the tropical forests are disappearing at an astounding rate that will see them completely removed from the face of the earth in at most three hundred years—assuming that the rate of their removal does not increase. This latter assumption requires that human population size does not grow.

Perhaps the most balanced treatment concerning forest loss comes from biologist/paleontologist Daniel Simberloff of Florida State University. Simberloff is a quantitative ecologist. Along with paleontologists David Raup and Stephen Jay Gould, he has been involved in formulating mathematical models for past species diversity. He has applied mathematical models to try to estimate present and future extinction rates in the tropical forests. In a 1986 article, Simberloff analyzed all available information regarding the rate of forest destruction, data mainly derived from satellite imagery and remote sensing. He found that tropical forests in Asia are already virtually gone and will be completely extirpated by the year 2000. Concentrating on American tropics, he found that even using the most conservative measures, 40 percent of the original area of forest (the area of tropical forest in the Americas at the beginning of this century) will have been destroyed by the year 2000. The area of parks, reserves, and protected tropical forest in all of the Americas amounts to about 60,000 square miles. At current deforestation rates, this is approximately the area of rain forest destroyed every two years.

To compute rates of species loss, Simberloff utilized the work of two famous ecologists, Robert MacArthur and Edward O. Wilson, who in the 1960s formulated a theory of species diversity called the equilibrium theory of island biogeography. This theory relates the area of habitat to the number of species present; as habitat area increases, so too do species numbers, and they do so in a predictable way. Similarly, as habitat area decreases, species numbers likewise fall. Because the number of species bears a predictable relationship to area available, deforestation leads to a shrinking of habitat in a way amenable to analysis. MacArthur and Wilson's equations can be used to predict rates of extinction. But their studies showed an even more alarming result.

In their studies on the number of species present on islands, they found that, for equal areas, an island always has *fewer* species than a mainland or continental area of similar size, even if the habitats are otherwise exactly identical. The implications of this are frightening. It means that parks and reserves, which essentially become islands surrounded by disturbed habitat, will always suffer a loss of species. It also means that cutting up the rain forest (or any forest) into patches of disturbed and undisturbed regions, creating many "islands" of forest, will greatly increase the rate of extinction.

Simberloff used MacArthur and Wilson's mathematical conventions to predict rates of species loss among two groups of American tropical organisms: birds and plants. There are currently about 92,000 recognized plant species (and an unknown number waiting to be described by science) in the New World tropical rain forests, and 704 species of birds. Simberloff calculated that by the year 2000, almost 14,000 plant species (15 percent of the total) and 86 bird species (12 percent of the total) will have gone extinct, assuming the most conservative estimates of extinction rate; these are minimum numbers. Missing from this calculation is the number of species of plants and birds that has already gone extinct since mankind's arrival in the New World.

Even more frightening numbers emerge if we look slightly further into the future. If the tropical forests become restricted to current and planned reserves and national parks, Simberloff predicts an extinction of more than 60,000 plant species (66 percent) and 487 bird species (69 percent). Again, these are minimum numbers. At the current rate of deforestation, these extinctions can be predicted to occur somewhere between 2050 and 2100. Simberloff concluded his article with the following sentence: "The imminent catastrophe in tropical forests is commensurate with all the great mass extinctions except for that at the end of the Permian." But even that dire statement is not quite right. Although the *percentage* of extinction (the relative numbers of the total fauna going extinct, as predicted by Simberloff) is less than that of the First Event, the *absolute* number of species going extinct in the tropics during the Third Event will be far greater, simply because the diversity of species is so much higher in today's world. Wilson has most recently, and perhaps most cogently, summarized the decline in biodiversity. He estimates that *currently*, about 27,000 species undergo extinction each year in the tropical rain forests. By the year 2022, Wilson considers that a 20 percent extinction in total global diversity is a "strong possibility" unless the current rate of environmental destruction is slowed.

One final and great difference exists between the Third Event and all previous extinctions: This is the first known extinction in which large numbers of plant species are going extinct. Paleobotanist Andrew Knoll of Harvard University, a member of the National Academy of Science, has noted a curious fact: In all previous mass extinctions, plant species have proven remarkably extinction resistant. In both the First and Second events, animal species disappeared at far higher rates than plants. The current and impending extinction of plants thus renders the Third

Event unique, and ultimately very frightening. When plants go extinct, animal species soon follow.

4

Of the many statistics relating to the current extinction, one group of numbers transcends all others: the size of the human population. Ultimately, the fundamental cause of the Third Event is the fact that there are enormous numbers of people presently on the earth and many more are on the way.

The growth of the earth's human population shows an interesting shape when graphed; it appears to be a logistic, or *S* curve, with a long slow rise, followed by a steep rise in the middle, and then (it is hoped) a flattening out. World population grew relatively slowly for the first 100,000 years of human existence; it took our species (assuming that *Homo sapiens* of the group inhabiting the earth today first evolved 100,000 years ago) that long to produce a population of 1 billion individuals (a milestone reached about the year A.D. 1800). From that point on, however, the number of humans on the earth began to multiply very rapidly indeed, as we entered the steep growth portion of the curve. It took 130 years to reach the second billion, thirty more years to reach 3 billion, fifteen more to reach 4 billion, and an additional twelve years to reach 5 billion. By 1992 the population reached 5.5 billion. One billion of that population currently lives in utter, abject poverty. Nevertheless, the growth curve will continue to rise throughout this century and through much of the next before the leveling-off portion begins. The peak human population is expected to occur either late in the twenty-first century or early in the twenty-second; if current efforts to reduce birthrates succeed, the maximum population will be between 10 and 11 billion people, or about double today's level; if, however, birthrates continue at their current level, the human population will theoretically rise to 15 billion people. Barring nuclear war, these numbers are immutable.

The effect of all of these people on the earth's biota can be described using another set of numbers. Biologists have defined a measure of our planet's productivity, called net primary production, or NPP. This is an estimate of the amount of energy that green plants bind into living tissue through photosynthesis, minus the amount of energy the plants use to fuel their own life processes. Paul Ehrlich of Stanford University has recently summarized work by Peter Vitousek and others concerning the way in which humans utilize the world's organic production. It has been estimated that 225 billion metric tons of organic matter are produced annually, 60 percent on land and 40 percent in the sea. Humanity is currently using 3 percent of the world total (4 percent of the land total) each year, as food, firewood, or feed for livestock. But these direct consumptions of NPP pale compared to indirect usage. If land clearing, the parts of plants grown but not eaten by humans, and the parts of pastures maintained but not directly consumed by livestock also are taken into account, it turns out that humanity is currently co-opting 30 percent of the NPP. By the middle of the next century, this may rise to 80 percent, based on even conservative population increases. Our planet cannot withstand such numbers. To realize enough organic productivity, virtually the entire arable land surface of the earth—every forest, every valley, every bit of land surface capable of sustaining plant life, as well as much of the plankton in the sea—will have to be turned over to crops, if our species is to avert unprecedented global famine. In such a world, animals and plants not directly necessary for our existence will probably be a luxury not affordable. Those creatures that can survive in the vast fields and orchards will survive. Those that need virgin forest, or undisturbed habitat of any sort, will not.

Two further aspects of population growth deserve mention. The first relates to energy consumption. Although fossil fuels currently account for much of the energy needs of North America, Europe, and parts of Asia, the peoples of the more tropical regions of our world depend largely on firewood. The search for firewood, as much as any other factor, will necessarily contribute to increasing rates of deforestation as population numbers rise.

A second major issue currently facing us is overgrazing. A United Nations study issued in March 1992 reports that more than 10 percent of the world's most productive farming regions has been seriously damaged and rendered less fertile, or infertile, since the end of World War II. The area affected is as large as India and China combined, and is largely the result of continued overgrazing by livestock or of unsuitable agricultural practices, such as excess fertilizing. The three-year study involved 250 soil scientists, commissioned by the United Nations to monitor and report on the status of farmlands from around the world. The actual damage results in heavy erosion, which has removed topsoil. Two-thirds of the ruined land is in Africa and Asia. The report also concluded that the per-capita food production in eighty of the developing countries surveyed has declined over the past decade.

All of these factors suggest one conclusion: A lot of hungry humans are going to be vying for the world's food resources during the next century. And with so many hungry *Homo sapiens* running around, the game preserves, sanctuaries, and various national parks set aside to preserve wildlife are going to be increasingly tempting targets for poachers, squatters, and just plain hungry people. Maintaining the integrity of the forest and nature reserves, themselves the greatest hope for staving off unprecedented levels of species extinction, may be the greatest challenge of the twenty-first century.

5

Another set of numbers will ultimately have much to do with the eventual severity of the current extinction: the changing levels of critical gases in our atmosphere. Ironically, it is not the major two atmospheric constituents—nitrogen and oxygen—that are changing, but two gases found in relatively minute quantities: carbon dioxide and ozone. One is increasing in concentration, the other decreasing. The result of these minute changes may spell disaster for many species currently living on the earth.

As far as anyone can tell, we are still in the Ice Age. The pattern for the last 2 million years has been one of prolonged cold, lasting about 100,000 years, followed by much shorter warm periods, or interglacials. The current warm phase began about 18,000 years ago and continues through the present day. According to past patterns, our world should begin a new cooling period, perhaps with renewed continental glaciation, some time in the next few thousand years. Human activities, however, may have changed that scenario. Our species has changed and perturbed not only the surface of the earth and its oceans, but the very nature of the atmosphere as well.

Our planet stays warm in large part because of a phenomenon known as the greenhouse effect. Some sunlight passing through the atmosphere

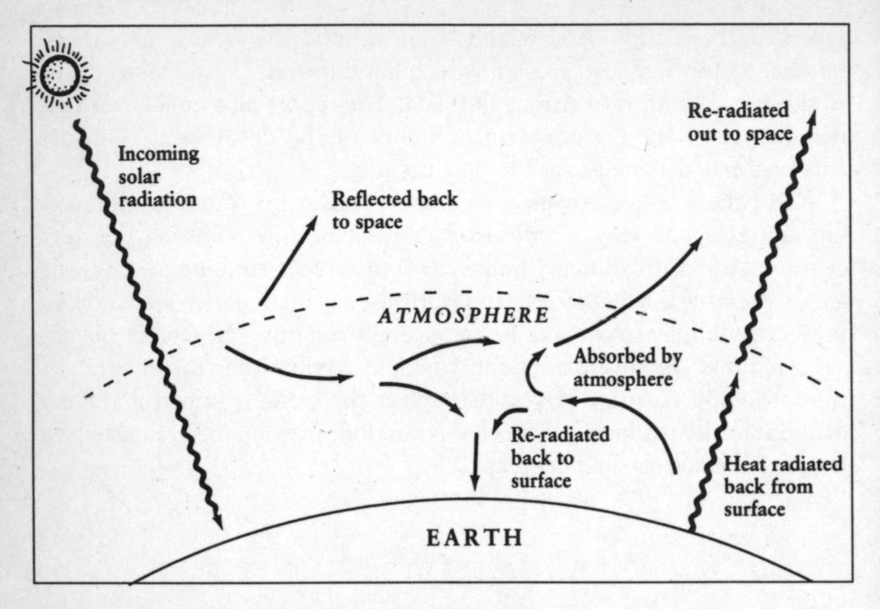

Diagram showing how the greenhouse effect takes place. About 40 percent of incoming solar radiation is reflected back into space. The warm surface of the earth radiates heat back into space at infrared wavelengths. As greenhouse gases increase in the atmosphere, however, some of this radiation is absorbed by the atmosphere and re-radiated back to the earth's surface.

hits the earth and is reflected back upward. Some escapes back into space; much of the reflected energy, however, is absorbed by particular gases that are transparent to sunlight but opaque to longer-wavelength radiation such as heat. The most important greenhouse gas is water vapor. Another is carbon dioxide.

Carbon dioxide is an extremely important gas for life on the earth. Plants require CO_2 to complete photosynthesis, a process that releases oxygen into the atmosphere. But too much CO_2 increases global warming by increasing the greenhouse effect. The amount of CO_2 in the atmosphere varies during the glacial cycle; during the cool, glacial intervals the atmosphere contains about 0.020 percent CO_2. During the warmer, interglacial periods, the CO_2 value rises, to about 0.028 percent. Today the level is 0.034 percent and rising. The amount of CO_2 in the atmosphere has risen 25 percent since the year 1800 and is expected to double sometime during the next century.

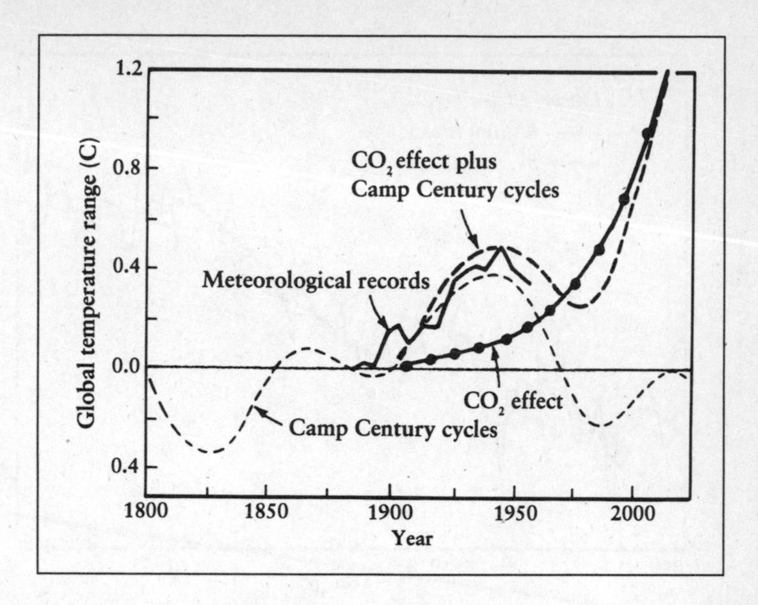

Prediction about global temperature rise was first published in 1975 by atmospheric scientist Wallace Broecker. It was one of the first predictions of greenhouse effect global warming. Temperature variations since 1975 have closely matched this forecast.

The increasing amount of CO_2 is coming from several sources, the most important being the burning of fossil fuels. But another source is deforestation. When forest areas are cleared by burning, vast quantities of CO_2 are released. But a more insidious effect of forest clearing is to lower plant productivity in the area because croplands almost never support as much plant tissue volume as do forests and use less CO_2. As the level of CO_2 continues to rise in our atmosphere, thus not allowing heat to be able to escape into space, the result is that our planet is warming rapidly.

Global warming does not sound like such a bad deal if you are stuck in the middle of a New England winter. However, virtually everyone agrees that the long-term consequences of even slight increases in average temperature values for the globe as a whole are potentially devastating. If humanity is to stave off global famine on an unprecedented scale, the major crop-producing areas must maintain a constant and predictable supply of food. Global warming will reduce productivity in many

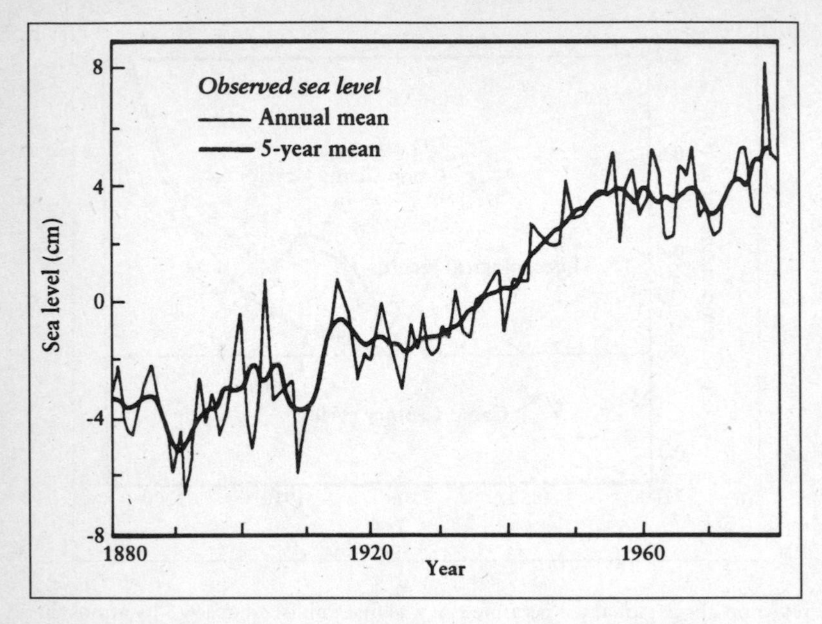

Rising trend in global sea level, superimposed on rising global temperature.

of the tropical and subtropical regions, where the growing seasons are longest. Global warming will increase the spread of deserts, and it will increase droughts and reduce water supplies in some areas of the globe, while greatly increasing precipitation in other regions. At a minimum, vast regions of the earth will experience rapid climate change, with dry areas becoming wetter and many wet areas becoming drier. Some regions will benefit greatly, both from increasing precipitation and from the increased amount of CO_2 itself, for plants have been shown to grow faster if given excess CO_2. But many other areas will become less conducive to life. If any lesson emerges from the study of past extinctions, it is that rapid environmental changes lead to species death. Whatever its ultimate effects, the pulse of global warming we have embarked on is sure to lead to massive environmental changes over much of the earth.

Atmospheric scientists have determined that the average temperature of the earth has increased by about 1°C since 1860. This change seems insignificant, but it has occurred very quickly: The predicted temperature increases will be fifteen to forty times faster than any temperature change ever experienced on the earth. No one can say what this may do to the planet's biota. But most animals and plants have very narrow temperature and moisture requirements. When climate changes slowly, climate belts slowly move northward or southward, and animals and plants often can migrate to more acceptable regions. Atmospheric scientists now predict that the world will warm at a rate of about 0.5° C each decade. This rate of warming will shift climate zones so rapidly that forests will have to move at a rate of 400 miles each century to stay within their required temperature and rainfall belts, which is not possible. It remains to be seen if plant and animal species can migrate successfully out of new conditions this rapidly. At the end of the last Ice Age, changing climate zones caused spruce forests to migrate northward at a rate of about 100 miles per century, a rate considered something of an arboreal speed record. Forest scientists are already predicting the start of a temperate forest "dieback" in North America by the late 1990s. Forests experiencing dieback will be full of aging, mature trees but without new saplings of the same species. Forests will eventually be replaced by weedy species. If plant species cannot migrate quickly enough, they will go extinct.

Predicting the consequences of global warming is enormously complex and can be done only with computer modeling. Because of the many variables and assumptions fed into these models, there are no certain outcomes. Today meteorologists find it almost impossible to predict the weather more than a few days in advance, because many atmospheric phenomena exhibit chaotic rather than predictable behavior; the entire new science known as chaos theory was originally derived from studies of the atmosphere. Imagine, then, attempting to predict weather patterns in a future, warming world. Yet, even if detailed understanding of global warming still defies prediction, certain generalities emerge. Virtually all scientists agree on one effect of global warming: The seas are going to rise.

Sea level is tremendously dependent on the mean global temperature and the extent of the ice caps. During colder periods, when a great deal of seawater is locked up in glaciers and ice caps, sea level is low. In warmer periods, when there is little ice, the levels of the seas are much higher. Only 18,000 years ago, when the continental glaciers of the last Ice Age were at their maximum extent, the level of the sea was more than 300 feet lower than it is now. Because of this, Asia and America were joined by a land bridge, and Australia was joined to Tasmania and New Guinea. If the globe continues to warm, the great Greenland and Antarctic ice sheets may begin to melt, and, as they do, the level of the seas will rise rapidly. The rate and extent of that sea level rise will have a great deal to do with the fate of mankind's billions and, therefore, with the fate of the many millions of species currently in the stewardship of humanity.

Like so much else about the global warming issue, there is great uncertainty about how fast the seas will rise, and how much. In 1990 atmospheric scientists estimated that by the year 2050, when carbon dioxide levels will have doubled from today's values, the seas will have risen by between 1.5 and 15 feet above their present-day levels. Newer estimates have reduced those figures somewhat, because global warming may increase precipitation in the polar cap regions and thus cause snow accumulation to keep pace with the melting. At a minimum, however, thermal expansion of the seas will cause at least a one-foot rise early in the next century. Most scientists involved in the issue seem to think that at least a *two*-foot rise by 2030 is inevitable. But this figure itself is somewhat of a minimum estimate; most models assume that the rise in atmospheric CO_2 levels will cease after doubling, and that may be wishful thinking. As the world's population increases, more and more fossil fuel and wood will be burned to keep people warm and nourished, and hence the level of CO_2 in the atmosphere may continue to rise even after it is twice the current level. If so, it will cause the earth to warm even further and the seas to continue to rise. Currently more than 1 billion people live in areas that would be inundated by the sea following a fifteen-foot sea level rise. These low areas are also regions of great agricultural importance; for instance, most of the world's rich deltaic regions, such as the Nile, Mekong, Mississippi, and Rhine river deltas, would be inundated. In addition to the actual land lost, a rising sea level would change water tables and cause saltwater to be injected into currently productive farmlands. A recent report issued by the Netherlands (a country with a lot to worry about) asserts that about 3 percent of the world's current land area would be either covered by the sea or rendered infertile due to the intrusion of saltwater if current rates of sea level rise continue through the next century. This area includes one-third of the world's croplands. The rise in sea level would coincide with the period of maximum human population. The scenario conjures up visions of human displacement on an unprecedented scale, where "environmental refugees" by the millions or billions try to find safe haven and farmland. In such a world, will anyone worry or care about the sanctity of game preserves or the protection of endangered species?

Rising sea level poses many threats. Oceanographers estimate that the rise of one inch in sea level results, on average, in the erosion and removal of one foot of beach; for each foot of sea level rise, the boundary between salt- and freshwater at river mouths moves upstream by nearly a mile. The cost of containing the rising sea level will be staggering. In the United States alone, the computed cost of protecting currently developed coastal areas by the year 2100 is projected to be between $73 billion and $111 billion, at 1988 prices, and even then a land area the size of the State of Massachusetts will be lost to the sea. Other low countries, such as the Netherlands, Bangladesh, and many Pacific island groups, will lose huge areas of land, unless extremely expensive dikes are constructed.

The worst-case scenario would occur if global temperature were to rise by 10° C. Such a temperature change would cause all of the polar ice caps to melt. The seas would rise by 240 feet, and for the first time since the Age of Dinosaurs, large areas of North America would be inundated by a huge inland sea.

6

The rise in atmospheric carbon dioxide and other greenhouse gases will surely have major effects on the world. Ironically, it is the *removal* of another low-volume gas, ozone, that may create equally disastrous ramifications for the biota.

The thinning of the world's ozone layer was first detected in 1985 by a team of British meteorologists monitoring the atmosphere over Antarctica. They detected a 40 percent drop in stratospheric ozone from 1960s baseline levels. Most scientists greeted this initial report with some skepticism. But this astonishing discovery was soon confirmed. It seems that for several months each year, a hole opens in the ozone layer in the southern hemisphere. Ozone is a form of oxygen, and it serves a vitally important function for the world's creatures: It blocks much of the sun's damaging ultraviolet radiation from striking the earth's sur-

Rain forest, Hawaii.

face. Scientists worked furiously to figure out why a thinning and disappearance of the ozone was suddenly occurring.

The answer to the ozone hole mystery was supplied by atmospheric chemists Sherwood Rowland and Mario Molina. These two scientists found that the ozone was being destroyed by man-made compounds leaking into the atmosphere. The culprits were a group of chemicals called chlorofluorocarbons, or CFCs. These industrially produced chemicals found their way into the stratosphere, broke down into their component parts, and then destroyed ozone molecules.

In 1991 evidence showed that, for the first time, ozone holes were being produced in areas other than Antarctica; researchers discovered a large ozone hole over Europe. In the same year other scientists found the first unmistakable evidence that excess radiation, striking the earth beneath the Antarctic ozone hole, was reducing the productivity of oceanic plankton. The implications are ominous. If the ozone continues to be depleted, increasing amounts of ultraviolet radiation will bombard the earth continuously. Agricultural scientists have already tested some two hundred species of land plants with regard to increased ultraviolet exposure. Soybeans, beans, peas, squash, and melons are all susceptible to increased amounts of ultraviolet, and show reduced leaf sizes, truncated growth, poor seed quality, and increased vulnerability to disease as a consequence. As yet, no one knows what increased ultraviolet radiation will do to individual animals other than humans. In humans, the result is higher rates of cancer; can we assume that increased rates of ultraviolet radiation will be any less catastrophic to other groups of animals? The message is abundantly clear. Holes in the ozone will decrease crop productivity and put many species at greater risk of extinction.

7

The onset of the Third Event has long been under way. Yet all that has died to date may be but a prelude to the grim age of death about to begin.

Should we care? I often read rationalizations about why species should be preserved. All too often these polemics center on future benefits to mankind, such as new foods, better medicines, panaceas of all kinds derived from animals and plants known and yet to be described:

The message is that we should save species because it is in *our* species' interest to do so. But isn't there a question of ethics involved? Our species, *Homo sapiens,* is one of the newest on the earth. What right do we have to drive other, older species into extinction? Terry Erwin, the man who discovered how poorly we have done in estimating world species diversity, made a point in a recent *Science* article that there is far more at stake than food potential, or medicinal herbs, or other *economic* potential. He discusses species in terms of their *evolutionary* potential. Erwin stresses that we must not preserve just for the value of a species today, but because we have no idea what that species and its descendants might become. Who is to say that our species will be the last as well as first species on the earth to develop intelligence? Who can say that some currently insignificant species might not be the rootstock of some far-flung intelligence ultimately of greater achievement, wisdom, and insight than our own? Who would have predicted that the first protomammals migrating into the frigid Karroo 275 million years ago would give rise to the mammals, or that the small arboreal mammals trembling in fear of the mighty dinosaurs 75 million years ago would one day give rise to us? Who knows what the spotted owl may become, or what great societies may ultimately rise from the small beetles now living in the vanishing rain forests? And even those species destined to remain but small links in the ecosystem—each is the result of a great, long history, and each has some ancestor that has weathered great past disasters. What right do we have to kill them?

Eucalyptus trees are among the most beautiful trees to have ever evolved, in my opinion. Along the sunny coastlines of Africa, Australia, and California, they tower over the landscape, slim and graceful. But their silent beauty masks a quiet destructiveness, for they are among the most monstrous organisms on the earth. Outside of their native Australia, their leaves and bark are so toxic that they kill all plants around them and ensure that there will be no competition. The eucalyptus are not knowingly malicious creatures; like our species, they did not evolve with evil intent. But in their grace, beauty, and utter destructiveness they are almost human.

8

During the year it took to write this book, I gave numerous lectures and interviews about extinctions. I was always asked about the prospects of saving species, and always I felt uncomfortable with this question. I am a historian, spending my time long in the past; my time machine does not travel into the future. I can only parrot what I read about conservation and preserving what is left. Yet it seems incontrovertible that to save a significant portion of current biodiversity, we need to set aside gigantic natural reserves, places where humans are not welcome, places that will exist for millennia, places that will weather the upcoming human population explosion. There is great hope, because in the eleventh hour, much of the industrialized world, at least, has recognized that a giant ecological disaster faces us. In the United States, Japan, and Europe, great strides are being made, and the large reserves being set aside in these and other countries will ensure that much will be saved. But it is in the tropics that the greatest crisis looms. Can even a quarter of the rain forests and their myriad, enclosed species be preserved? Most likely this will not happen. Perhaps the best we can do is save as much as we can, and take note of the rest; perhaps the best we can do is make a record of what there once was, prior to the Third Event.

There is consolation. Who knows what great empires will rise from the ashes of the Third Event, some 5 or 10 million years from now? After the current winter will come a new spring, just as after every former mass extinction, genesis has proceeded anew.

9

I have a son. He is tall and gangly, with a face speckled by a galaxy of freckles. He is mischievous and playful, willful and happy, the normal mix of boyish hopes, dreams, and emotions. He is precious to me beyond belief.

I keep having this vision, of living with him in the Amazon rain forest, where we exist in a small hovel no different from that inhabited by a fifth of humanity. And in this dream, my son is hungry. Behind our house sits one last patch of forest, and in that pristine copse is the nest of a beautiful bird, the last nest, it so happens, of that species. This vision is a nightmare to me, because even knowing that these birds are the last of their race, I don't have the slightest doubt what my actions would be: To feed my son, to keep him alive, I would do whatever I had to do, including destroying the last of another species.

Anyone who thinks he or she might do otherwise is probably not a parent. There are a great number of parents currently on the earth, and many more on the way.

Chapter Twelve

Hope

I

I dropped my hammer to the ground, shrugged off the heavy pack, and then wearily sat beside the equipment on the summit of the grassy hilltop. The vast panorama of the surrounding Tunisian hills was spread out before me, but I concentrated on the distant knot of people on the next hillside. From my high vantage point, they looked like tiny swarming dots, an insect assemblage perhaps; creatures busily involved in some earnest but seemingly unfathomable activity so characteristic of ants and people. But I knew well what transpired on the next hill. Occasionally a reflected glint of light flashed my way, for the people there were wielding hard iron against the African earth. There was no shade atop either hill, and drops of briny sweat began to fall from my forehead, to be instantly swallowed by the dry earth around me. I looked again at the distant workers. In that group were some of the best-known geologists in the world, men and women from far-distant lands assembled here to study the world's best-preserved Cretaceous-Tertiary boundary site. I knew that the heat was ferocious on the far hill where they worked, but even so the furious activity was unrelenting. I thought of the famous saying about mad dogs, Englishmen, and noonday sun, silently appending geologists to the list. I admired their persistence and raged at their myopic view of the Second Event.

I had come to Tunisia in the spring of 1992, after completing (or so I thought) the last chapters of this book. I had come to resume my work, researching the causes and consequences of the Second Event, but also to escape from the terrible implications posed by the current extinction and from the nagging depression that my research and writing had produced. My research had created within me a black fugue, a feeling that there was little hope for the present-day biota in the building crescendo of the Third Event and little that any individual human could do to help derail the oncoming catastrophe. I did not fear for our species—to the contrary, I believed then and believe now that our brains and technology will allow us to survive anything the earth can throw at us—but at what cost to our fellow species? The answer to that question seemed too clear, and consequently I had come to Tunisia to bury my head in 65-million-year-old sand.

It seemed fitting to return to Africa, where I had begun this book a year earlier. Far to the south of where I now sat lay the Karroo, with its weathering bones of the First Event's victims. Around me, on this day, lay the skeletons of much younger victims, for the Tunisian hills are largely composed of chalks and microfossil oozes, made up of species decimated in the Second Event. And in between these two ancient graveyards, one on the southern tip of Africa, the other on the northern end, lay the entire Dark Continent, a great, modern cemetery rapidly filling with victims of the current extinction. I had come full circle, and the circle seemed closed.

As soon as the Tunisian conference on the Cretaceous-Tertiary extinction began, I quickly recognized that my geological colleagues were obsessed with the thin layer of clay marking the Cretaceous-Tertiary boundary and with little else. The conference had originally been arranged for 1991 but had been postponed due to the Gulf War. Our goal was to conduct a blind sampling program of the KT boundary in order to observe—in an impartial way—how rapidly microfossil species went extinct there. At our initial meeting, held in a venerable hotel in Tunis, both proponents and opponents of the meteoric theory unleashed their salvoes in spirited debate. But only geologist Gerta Keller and I proposed that our sampling should begin in strata deposited at least 2 million years prior to the thin clay layer with its evidence of meteoric catastrophe, rather than in strata only 2,000 years older. Our arguments seemed to fall on deaf ears, for even those arguing against a meteoric cause of the Second Event seemed unwilling to view the extinction as being multicausal. And so when we finally arrived at the field site near the ancient village of El Kef, long ago designated by a ruling body of geologists as the world's standard reference section for studying the KT boundary, I was left to my own devices and to my own sampling. Over several days I sampled strata deposited during the last 2 million years of the Mesozoic Era, looking for and finding evidence that the Second Event was indeed the product of many things: climate change, sea level change, and ultimately a meteor strike—just as the current event stems from many causes.

My musings were interrupted by a gentle tug at my sleeve. A small Arab girl shyly stood beside me, a cup of strong mint tea being proffered.

"M'sieur?"

I looked at the girl, trying to judge her age. She wore a long skirt and had beautiful if tangled hair. I took the cup and smiled, causing her to giggle and run off to the tiny knot of girls all breathlessly watching among their sheep. The tea was hot and strong, a gift from the nearby village; I drank it gratefully, its strong caffeine and sugar giving an immediate burst of energy. What must we look like to them, I thought to myself, with our uncovered women working alongside the men, our large trucks, our plentiful food, but above all our unfathomable tasks, our digging in the rocks to grub out small bits of stone, our brutally hard work not for food, or clothing, or any necessity, but simply to put handfuls of rock into white cloth bags, and our endless arguing among ourselves all during the process.

Finishing the tea in a gulp, I stood up and stared outward, once again, into the foothills of the Atlas Mountains, looking south into the hazy heat. It was hard for me to imagine this place as Africa. On this, my first trip to northern Africa, I had expected to see a wilderness of some sort, a desert, perhaps, replete with camels and wild beasts, exotic lizards and palm-fringed oases. Instead I found a land so long cultivated by humans that all vestiges of its original heritage had been long erased. The chance play of mountainous topography and moisture-laden air from the nearby Mediterranean Sea had made Tunisia a rich, rainy enclave along the northern African shore, an incongruous green emerald of fecundity and fertility set against the dry parchment of the Sahara. Since the end of the Ice Age, successive conquerors had changed the original Tunisian grasslands into tamed fields of wheat, enriching succeeding empires in the process. The Romans had built great aqueducts from the rain-rich areas of the coast into the dry interior and, in bringing plentiful water into these regions, had pushed the great desert back even farther. More species lived here now than at any other time before the arrival of mankind to this region. Human presence has been unwelcome in many places on the earth, such as the rain forests. But people have also transformed lifeless regions for the better, creating havens that allow life to blossom.

Lengthening shadows proclaimed the swift passing of the day, and it was time for me to resume my sampling. I shouldered my pack and looked once more to the south, into the heartland of Africa. Far off in the distance I could barely see the brown hills of the Sahara. But between me and the lifeless desert there lay a green land. For tens of miles the green fields seemed to hold the advancing desert at bay, a dike of life put there by human hands. Perhaps it was the welcome tea, or perhaps only the passage of time. But for whatever reason, from that time I began to have hope.

2

The greatest hope for the species of our world comes from the expanding realization of how real the danger of mass extinction is. Unprecedented events occurred in 1992, with the most hopeful being the environmental summit held in Rio de Janeiro. Virtually alone among nations the United States dragged its feet, and I would like to think that the electoral defeat of George Bush in 1992 stemmed in no little way from his administration's refusal to acknowledge the deepening ecological crisis facing the earth, both at the conference and later, in the United States during the presidential campaign.

Politicians in many countries are finding that they can win elections by promoting themes of conservation. Some countries containing tropical rain forests—and the great arkload of species inhabiting them—have begun ambitious projects to save species. Of all such examples, perhaps Costa Rica in Central America offers the most hope. Costa Rica is the most stable democracy in Central America. It has a long history of conservation, and has recently begun a survey in an effort to catalog the richness of its animals and plants. Approximately 20 percent of the land area of Costa Rica is now protected. On the other hand, Costa Rica suffers from many of the problems of other developing countries, and these problems all impact on the region's biodiversity. Costa Rica has one of the highest birthrates in all of the Americas, and a large foreign debt. About 95 percent of the country was originally covered with forest, but a great amount deforestation has occurred as forests are replaced by cattle ranches. Within ten to fifteen years, only the parks and reserves will still contain forest.

Paradoxically, perhaps the single greatest hope for the animals and plants of Costa Rica, and for many other tropical countries as well, is tourism. Poorer countries are finding that nature reserves and game parks can make more money than farmland. This tenet was brought home to me on a chill October night in Seattle by Richard Leakey, son of the famous paleoanthropologists Louis and Mary Leakey and a noted paleoanthropologist in his own right. I had the privilege of introducing Richard to a packed house on my campus, and I assumed that most of the audience had come to hear of his work on uncovering our origins and ancestors. But Leakey has had a change of profession recently, and I was surprised when he spoke of conservation, not paleontology. No longer does he prowl the ancient sediments of East Africa's Great Rift Valley in search of hominid fossils. Now he is minister in charge of conservation for the nation of Kenya. His major concern over the past several years has been in slowing or halting the ivory trade, a practice that has led to the slaughter of great populations of African elephants. Leakey's methods are direct: He maims or executes those caught killing elephants for ivory. For this offense, he must now travel in his native Kenya with bodyguards. Using violence to promote conservation seems counterproductive to me, but Leakey is the man on the spot. He hopes that increasing numbers of people will travel to the great game parks in Africa to see what is left, and in so doing help protect what is left. Leakey promotes ecotourism; if enough currency comes into Kenya, then the great game parks will be preserved. As Leakey spoke I felt an eerie chill, for here was a man who had spent a life much like my own, prowling the outlands of our planet for scraps of past life. Yet he had renounced this profession, and did so again publicly to his Seattle audience—to dedicate his life to saving what is left, to help turn back the tide of the Third Event. Very few of us can change our professions in midlife. But all of us can take tiny steps that, in concert, can have enormous effect on the diversity of life on this planet. We can use less paper products, selectively refuse to buy products made of tropical woods, use less energy, elect politicians willing to back conservation efforts . . . the steps are almost intuitive.

By coincidence, two nights after Richard Leakey's visit to Seattle another conservationist addressed an audience—perhaps largely the same one—at my university. Norman Myers was the speaker on this night, and I had a chance to see the man who first highlighted the plight of the rain forests and who first called attention to the tide of extinctions occurring in the tropics due to their destruction. Myers's message was not so dissimilar to Leakey's: There is still time to take action, still time to slow the slaughter, still time to save the majority of species living on our planet. But Norman Myers did not mince words, or sugar-coat the medicine. He gives the tropical forests at most a decade of life. After that, clear-cutting will have removed them from the earth.

3

Long after delivering the first draft of this book to my publisher in New York, I received a copy of Edward O. Wilson's new masterpiece, *The Diversity of Life*. I was heartened (and at the same time saddened) at the similarity in our conclusions. Wilson has summarized the steps needed to take us back from the brink of further mass extinction. He promotes four goals:

1. *Survey the world's fauna and flora.* If we are to reduce the loss of biodiversity, we must first know what the level of biodiversity is and, even more important, better understand the biology of the threatened organisms and ecosystems. Wilson proposes to conduct biodiversity surveys at several different levels, from relatively rapid assessments in ecosystems particularly endangered, to more long-term studies.
2. *Create biological wealth.* Wilson argues that every country has

three forms of wealth: material, cultural, and biological. The latter wealth is what is being destroyed as extinction continues. Wilson talks of chemical prospecting, searching among wild animals and plants for new medicines and chemicals. If we come to realize the wealth of biodiversity, we will take greater care to ensure its survival.

3. *Promote sustainable development.* Wilson notes: "The proving ground of sustainable development will be the tropical rainforests. If forests can be saved in a manner that improves local economies, the biodiversity crisis will be dramatically eased." Gordon Orians of the University of Washington has pointed out to me that selective forestry in the tropical rain forests is a positive step; we cannot exclude our species from the tropics, and there must be economic development.
4. *Save what remains.* There is a tremendous effort under way by zoos, aquaria, and plant reserves to save endangered species from extinction. Unfortunately, most of these efforts are directed at large and attractive animals and plants. Wilson points out that these efforts will save only a tiny fraction of the creatures currently threatened by extinction. The greatest hope lies in saving entire ecosystems before they become endangered.

As I write this it is late November in my northern hemisphere city; the leaves are gone from the trees, and winter seems very close. Winter has already arrived in the Hell Creek region of Montana and in the far-off Caucasus Mountains as well, places where the bones of dinosaurs and other previous inhabitants of the world prior to the Second Event erode from the ages-old hills. In South Africa, by contrast, spring is well advanced, and in the Karroo great heat now grips the land. Yet the seasons, so precious and yet so swiftly passing in my own life, are of no import to the victims of the First and Second events. Those two great kingdoms are now the kingdoms of the dead. We have it in our power to avert or lessen the severity of a third such catastrophe, to ensure that our children's children will awaken to the sound of birds in a world still alive with species.

One of the founding principles of geology is called uniformitarianism. Geology's founding fathers realized the natural processes occurring on the earth today are a key to understanding the past. With regard to understanding the current crisis in biodiversity, perhaps the opposite is more appropriate. Those who do not believe that mass extinctions are enormous catastrophes would do well to study the past.

London, Cape Town,
Hawaii, Seattle,
1991–1992

NOTES

Introduction

Pages xi–xvi. My visit to the Philippine Islands took place during June and July 1987. I was based at Silliman University in Dumaguete City, Negros Island. Two weeks of trapping in the Tanon Strait resulted in only three captures, in areas that once supported a rich fishery. (See P. Ward, *In Search of Nautilus.* New York: Simon & Schuster, 1987.) Dynamite fishing and the even more catastrophic cyanide fishing (where cyanide was dumped into the coastal seas) was practiced in the Philippines between the mid-1960s and the early 1980s. Estimates about the drop in fisheries' yields come from Dr. Angel Alcala, now president, Silliman University, and from the Philippine Department of Fisheries.

Pages xvi–xviii. Numerous recent books and discussions describe and define mass extinctions. Perhaps the best recent summary is by Steven Stanley, *Extinction.* New York: Scientific American Books, 1987. For a brief history of mass extinction research, see Norman Newell, "Mass Extinctions, Illusions or Realities," Geological Society of America, Special Paper 190, 1982. For definitions of mass extinctions and their relative, see John Sepkoski, "Mass Extinctions in the Phanerozoic Oceans: A Review," Geological Society of America, Special Paper 190, 1982.

Page xviii. My designation of three "events" may be controversial to many colleagues, since most paleontologists agree that there were five "major" extinctions during the past 570 million years. (No one, however, disputes that the Permian and Cretaceous events were the most severe.) Three of the other Big Five extinctions either occurred before the evolution of land vertebrates or took place mainly in the seas; they were therefore of less importance to the history of vertebrates and, hence, to us. I have thus chosen to be deliberately anthropocentric. If a jellyfish was writing this book, it might define other extinctions than the ones I have chosen as the major events.

Chapter One

The Cape

I was the guest of the South African Museum during the latter part of 1991; I would like to thank my host, Dr. Herbert Klinger, and the director, Dr. Michael Cluver.

Pages 6–11. A more detailed discussion of the rise of metazoan life can be found in my recent book, *On Methuselah's Trail.* New York: W.H. Freeman, 1991, as well as in Steven Stanley, *Earth and Life Through Time,* 2d ed. New York: W.H. Freeman, 1989.

Pages 16–22. An excellent historical treatment on both the birth of biostratigraphy as well as John Phillips's contributions to understanding diversity and mass extinction through time can be found in Martin Rudwick, *The Meaning of Fossils: Episodes in the History of Paleontology.* New York: Neale Watson Academic Publications, 1976. For background on the intellectual climate among paleontologists during Phillips's time, see the excellent book by Adrian Desmond, *Archetypes and Ancestors, Paleontology in Victorian London, 1850–1875.* Chicago: University of Chicago Press, 1982.

Pages 22–23. Professor John Sepkoski's diversity results are found in several sources. The role of mass extinctions in regulating diversity is found in J. Sepkoski, "A Model of Phanerozoic Taxonomic Diversity," *Paleobiology* vol. 10, 1984, pp. 246–67.

Pages 27–30. Discussions and descriptions of Devonian life are found in Steven Stanley, *Earth and Life Through Time.*

Chapter Two

The Great Karroo

My visits to the Karroo were greatly aided by Dr. Roger Smith of the South African Museum. Roger spent endless hours patiently answering my questions about the Karroo and took me on extended tours of the Karroo Basin. I would also like to thank Dr. Gillian King and Dr. Michael Cluver for additional insight into Karroo geology and paleontology.

Pages 31–37. Background on Karroo geology was derived from many sources. An excellent summary is found in Nicholas Hotton, "Stratigraphy and Sedimentation in the Beaufort Series, South Africa," University of Kansas Department of Geology, Special Publication 2, 1967.

Pages 37–42. A summary of the theory of plate tectonics and the history of investigation can be found in many sources, but I recommend Steven Stanley, *Earth and Life Through Time.*

Pages 44–49. The history of protomammals in the South African Karroo desert and their anatomical descriptions are derived from Dr. Michael Cluver's illustrated book, *Fossil Reptiles of the South African Karroo.* Cape Town: South African Museum Press, 1991.

Page 49. An excellent history of the Boers and their epic treks is recounted in J. Omer-Cooper, *History of South Africa.* London: James Curry Ltd, 1987.

Page 50. The best description of Karroo sedimentary environments comes from Roger Smith, "A Review of Stratigraphy and Sedimentary Environments of the Karroo Basin of South Africa," *Journal of African Earth Sciences* vol. 10, 1990.

Pages 53–58. Much of the discussion of warm-blooded protomammals (and dinosaurs), as well as additional descriptions of the protomammals and their evolutionary history and mode of life, comes from Robert Bakker, *The Dinosaur Heresies.* London: Penguin Books, 1988.

Pages 58–59. I visited Lizard Island on the Australian Great Barrier Reef in 1985 and 1986. Both expeditions were concerned with capturing specimens of the chambered nautilus.

Page 63. Gould discusses change and evolution in his wonderful book, *Wonderful Life.* New York: Norton, 1989.

Chapter Three

End of an Era

Pages 65–67. Descriptions of the end-Permian extinction as well as a brief history of the Alvarez impact hypothesis can be found in Steven Stanley, *Extinction.* Details of the extinction in the Karroo Basin and specifically the purported gradual nature of the extinction among protomammals are found in many sources. Two of the most recent include Gillian King, "Dicynodonts and the end-Permian Event," *Paleontologia Africana* vol. 27, 1990, pp. 31–39, and Curt Teichert, "The End-Permian Extinction," in Earl Kauffman and Otto Walliser, eds., *Global Events in Earth History.* Berlin: Springer-Verlag, 1990.

Pages 67–68. The description of Permian stratigraphic sections and details of both the missing intervals contained therein as well as the history of investigation in China are found in Curt Teichert, "The End-Permian Extinction."

Page 70. I am indebted to Roger Smith and Gillian King for providing me much information about the vertebrate fossil holdings found in the South African Museum.

Page 76. The uncritically held assumption that some portion of Permian strata is missing from the Karroo Basin stems in part from Robert Broom, *The Mammallike Reptiles of South Africa and the Origin of Mammals,* London: H. Witherby Company, 1932, as well as a widely cited paper by J. Anderson and A. Cruickshank, "The Biostratigraphy of the Permian and Triassic: A Review of the Classification and Distribution of Permo-Triassic Tetrapods," *Paleontologia Africana* vol. 21, 1978, pp. 15–44.

Pages 77–82. The history of the various South African paleontologists comes from a chapter in *South African Men of Science*. Cape Town, 1986 and from recollections of various members of the South African Museum staff.

Pages 85–86. Permian isotopic results can be found in papers by Ken Hsu and J. McKenzie, "Carbon Isotope Anomalies at Era Boundaries: Global Catastrophes and Their Ultimate Cause," Geological Society of America, Special Paper 247 (1990), as well as in J. Thackeray et al., "Changes in Carbon Isotope Ratios in the Late Permian Recorded in Therapsid Tooth Apatite," *Nature* 347 (1990), pp. 751–53.

Page 86. The stratigraphic ranges of Karroo protomammals are found in A. Keyser and Roger Smith, "Vertebrate Biozonation of the Beaufort Group with Special Reference to the Western Karroo Basin," *Memoirs of the Geological Survey of South Africa* 12 (1979); and in J. Kitching, "Distribution of the Karroo Vertebrate Fauna," *Memoirs of the Bernard Price Institute of Paleontological Research* vol. 1, 1977, pp. 1–131.

Chapter Four

Dawn of the Mesozoic

Page 92. The curious and characteristic red color of Triassic rocks is due to the oxidation of iron sediments. A geological description of this phenomenon and its causes can be found in H. Blatt, G. Middleton, and R. Murray, *Origin of Sedimentary Rocks*. Englewood Cliffs, NJ: Prentice-Hall, 1972.

Page 94. Dr. David Jablonski of the University of Chicago has been one of the most articulate scientists unraveling the mysteries of extinction. His work on the relations between extinction and geographic range is summarized in "Causes and Consequences of Mass Extinctions: A Comparative Approach," in David Elliott, ed., *Dynamics of Extinction*, New York: John Wiley and Sons, 1986.

Pages 95–96. Descriptions and illustrations of *Lystrosaurus* and its world are found in Michael Cluver, *Fossil Reptiles of the South African Karroo*. Cape Town: South African Museum Press, 1991.

Page 96. The human flesh calculation assumes that there are 5.5 billion people, and that their average weight is about 75 to 100 pounds each. This calculation is thus a minimum figure.

Pages 97–98. The discussion of reptilian evolutionary pathways is drawn largely from Robert Carroll's epic tome, *Vertebrate Paleontology and Evolution*. New York: W.H. Freeman, 1988.

Pages 98–99. Triassic geological history is from Steven Stanley, *Earth and Life Through Time*.

Pages 99–101. A discussion of fossil footprints in the Connecticut River Valley is found in Edwin Colbert, *The Great Dinosaur Hunters and Their Discoveries*. London: Dover Publishing Company, 1984.

Pages 101–102. Much of the material for my excursion back into the ancient Petrified Forest of Arizona is derived from a book by Robert Long and Rose Houk, *Dawn of the Dinosaurs, the Triassic of Petrified Forest.* San Francisco: Petrified Museum Association, 1988.

Page 106. Details of the possible meteor impact occurring at the end of the Triassic Period comes from Paul Olsen and others, "The Triassic/Jurassic Boundary in Continental Rocks of Eastern North America," *Global Catastrophes in Earth History,* Geological Society of America, Special Paper 247. For an opposing view, the article in the same volume by Anthony Hallam, "The End-Triassic Extinction Event," arrives at a conclusion completely opposite to that of Olsen and his coworkers.

Chapter Five

The Age of Dinosaurs

Page 112. An excellent introduction to biomechanics can be found in Steven Wainright, *Biomechanics.* New York: Harvard University Press, 1988.

Pages 112–113. Details of the German East African expeditions and the discovery of *Brachiosaurus* are found in Edwin Colbert, *The Great Dinosaur Hunters and Their Discoveries.* London: Dover Publishing Company, 1984.

Pages 115–116. Ostrom's work on the dinosaur-bird transition can be found in John Ostrom, "Archaeopteryx and the Origin of Birds," *Biological Journal of the Linnean Society,* vol. 8, 1976.

Pages 115–116. The description of *Deinonychus* is found in Ostrom's article, "Osteology of *Deinonychus antirrhopus,* An Unusual Theropod from the Lower Cretaceous of Montana," *Bulletin of the Peabody Museum,* vol. 30, 1969.

Pages 117–118. Excellent summaries of the warm-vs.-cold-blooded controversies are found in Bakker, *The Dinosaur Heresies,* London: Penguin Books, 1988, as well as John Noble Wilford, *The Riddle of the Dinosaur.* New York: Alfred A. Knopf, 1985.

Chapter Six

Death of the Dinosaurs

Pages 124–128. The geology of the Hell Creek region is found in J. Gill and W. Cobban, "Stratigraphy and Geologic Age of the Montana Group and Equivalent Rocks, Montana, Wyoming and North and South Dakota," *United States Geological Survey Professional Paper* 776 (1973), and in David Fastovsky, "Paleoenvironments of Vertebrate-bearing Strata at the Cretaceous-Paleogene Transition, Eastern Montana and Western North Dakota," *Palaios* vol 2, 1987.

Pages 130–131. The various hypotheses concerning the extinction of the dinosaurs are nicely listed in John Wilford, *The Riddle of the Dinosaur.* New York: Alfred A. Knopf, 1985.

Page 133–134. McLaren's Presidential Address was published as D. McLaren, "Time, Life and Boundaries," *Journal of Paleontology,* vol. 44, 1970.

Pages 134–135. A discussion of Urey's discovery and how it was received is found in David Raup, *Extinction, Bad Genes or Bad Luck?* New York: W.W. Norton, 1991.

Pages 136–137. The history of the Alvarez discovery and other aspects of the controversy have been published by William Glen, "What Killed the Dinosaurs?" *American Scientist,* July–August 1990.

Page 137–138. The original Alvarez discovery was published by Luis Alvarez and others, "Extraterrestrial Cause for the Cretaceous-Tertiary Extinction," *Science* vol. 208, 1980.

Page 138. Smit's discovery was published as J. Smit and G. Klaver, "Sanidine Spherules at the Cretaceous-Tertiary Boundary Indicate a Large Impact Event," *Nature,* vol. 292, 1981.

Pages 140–142. The proceedings of the two Snowbird conferences were published by the Geological Society of America. Combined, these two thick tomes represent the best single source for information about extinction in general and the KT extinction in particular. The first was edited by L. Silver and P. Schultz and published as *Geological Implications of Large Asteroids and Comets on the Earth.* Geological Society of America, Special Paper 190, 1982. The second was edited by V. Sharpton and P. Ward, *Global Catastrophes in Earth History,* Geological Society of America, Special Paper 247, 1990. For brevity, these two volumes will be called Snowbird I and Snowbird II when referenced from now on.

Pages 144–145. The Hildebrand-Penfield exchanges are found in the June and November issues of *Natural History,* 1991.

Pages 146–147. The summary article by W. Alvarez and F. Asaro was published in *Scientific American,* October 1990.

Pages 149–150. W. Clemens, "Patterns of Extinction and Survival of the Terrestrial Biota During the Cretaceous/Tertiary Transition," in Snowbird I.

Pages 151–152. My work on ammonite extinction patterns was published as "The Extinction of the Ammonites," *Scientific American,* October 1983, and as "The Cretaceous-Tertiary Extinctions in the Marine Realm: A 1990 Perspective" in Snowbird II, and as "A Review of Maastrichtian Ammonite Ranges" in Snowbird II.

Page 152. The "Signor-Lipps effect" was originally described by P. Signor and J. Lipps, "Sampling Bias, Gradual Extinction Patterns, and Catastrophes in the Fossil Record," in Snowbird I.

Pages 152–153. The plant record at the KT boundary has been described by K. Johnson and L. Hickey, "Megafloral Change Across the Cretaceous-Tertiary Boundary in the northern Great Plains and Rocky Mountains, USA," in Snowbird II, and in A. Sweet, et al., "Palynofloral Response to KT Boundary Events: A Transitory Interruption within a Dynamic System," in Snowbird II.

Pages 154–155. The extinction of the large clams prior to the Cretaceous-Tertiary boundary is described by K. MacLeod and P. Ward, "Extinction Pattern of *Inoceramus* Based on Shell Fragment Biostratigraphy," in Snowbird II.

Pages 155–156. Deccan volcanism and its effect on the Cretaceous world is described by K. Caldeira and others, "Deccan Volcanism, Greenhouse Warming, and the Cretaceous-Tertiary Boundary," Snowbird II.

Pages 157–158. The vertebrate record across the KT boundary at Hell Creek is described by D. Archibald and L. Bryant, "Differential Cretaceous-Tertiary Extinctions of Nonmarine Vertebrates: Evidence from Northeastern Montana," Snowbird II, and in P. Sheehan and others, "Sudden Extinction of the Dinosaurs: Latest Cretaceous, Upper Great Plains, USA," *Science,* November 8, 1991.

Chapter Seven

Autumn

My trip to the Soviet Union took place in October and November 1990. I would like to thank the National Science Foundation and the Georgian Academy of Sciences for their sponsorship of this trip, and Jan Smit of the Free University, Amsterdam, for his expertise and fortitude. Anton Oleynik served as our host and guide in Moscow.

Pages 170–171. I have written of the rise of the angiosperms in greater detail in *On Methuselah's Trail.*

Page 173–174. The role of grasses and herbs in shaping the modern world is taken from Steven Stanley's two books, *Earth and Life Through Time* and *Extinction.*

Page 174. Details of the earliest mammalian radiations as well as the later history of mammals are found in Robert Carroll, *Vertebrate Paleontology and Evolution.* New York: W.H. Freeman, 1988.

Chapter Eight

Winter

Pages 178–179. Discussions of the Late Cenozoic history of the climate are found in Steven Stanley, *Earth and Life Through Time,* as well as Ken Hsu's article, "When the Mediterranean Dried Up," *American Scientist.*

Pages 179–182. The history of the primates comes from Robert Carroll, *Vertebrate Paleontology and Evolution,* and from Steven Stanley, *Earth and Life Through Time.*

Pages 182–183. My brother, Steven Ward, is currently professor of anthropology and chairman of the Department of Anatomy at Kent State University. The article reporting on the age of the earliest *Homo* was published by *Nature* (March 6, 1992).

Pages 185–189. Additional information about the Ice Ages can be found in E.C. Pielou, *After the Ice Age.* Chicago: University of Chicago Press, 1991.

Pages 189–191. A history of Cenozoic mammalian evolution comes from Carroll, *Vertebrate Paleontology and Evolution.* Dates on Ice Age events and history are from Stanley, op. cit.

Page 191–192. Loss of marine species is described in Steven Stanley, *Extinction.*

Chapter Nine

Overkill

Pages 194–196. The history of the Wenatchee Clovis dig comes from conversations with Dr. Don Grayson and Dr. Julie Stein of the University of Washington, from a public exhibit at the Wenatchee Historical Museum, and from an article published on March 10, 1992, in the *Wenatchee World* newspaper.

Pages 197–199. Much of the information about Overkill comes from a large volume, *Quaternary Extinctions,* edited by Paul Martin and Richard Klein (Tucson: University of Arizona Press, 1984). I derived information from chapters by D. Grayson; E. Anderson; Marcus and Berger; S. Webb; P. Gingerich; P. Martin; J. Mead and D. Meltzer; D. Steadman and P. Martin; R. Klein; R. Dewar; P. Murray; D. Horton; M. Trotter and B. McCulloch; R. Cassels; Olson and James; L. Marshall; and J. Diamond. I am especially indebted to Don Grayson for his time and insight in discussing these issues with me. Also important were Grayson's papers in the *Journal of Archaeological Science,* vol. 16, 1989, pp. 153–165, and the *Journal of World Prehistory,* vol. 5, 1991, pp. 193–231.

Pages 203–204. Information about vanishing North American wildlife comes from Roger Di Silvestro's book, *The Endangered Kingdom.* New York: John Wiley Co., 1989.

Chapter Ten

Lost Islands

Geological history of the Hawaiian Islands and much information about the various conservation problems there come from C. Stone and D. Stone, eds., *Conservation Biology in Hawaii,* Honolulu: University of Hawaii Press, 1989, and from Linda Cuddihy and Charles Stone, *Alteration of Native Hawaiian Vegetation.* Honolulu: University of Hawaii Press, 1990.

Pages 218–220. The story of humanity's arrival and early history in Hawaii comes from the sources just cited as well as from several works by Dr. P. Kirch: "The Chronology of Early Hawaiian Settlement," *Archeology and Physical Anthropology in Oceania* 9 (1974): 110–19; and *Feathered Gods and Fishhooks, An Introduction to Hawaiian Archeology and Prehistory.* Honolulu: University of Hawaii Press, 1985.

Pages 221–226. Information about nonnative species in Hawaii comes from W.

Gagne, "Terrestrial Invertebrates" in Stone and Stone, *Conservation Biology in Hawaii,* and from Cuddihy and Stone, *Alteration of Native Hawaiian Vegetation.*

Pages 226–231. The story of Hawaiian land snails comes from several sources, including long conversations with Dr. Rob Cowie of the Bishop Museum, Hawaii, and with Dr. Michael Hadfield, the University of Hawaii. Information about the destruction of land snails by collectors came from Michael Hadfield, "Extinction in Hawaiian Achatinelline Snails," *Malacologia* 27, 1986, pp. 67–81.

Pages 232–238. Information on the Hawaiian bird extinctions comes from conversations with S. Olson as well as from C. Stone, in Stone and Stone, *Conservation Biology in Hawaii,* and Olson and James, in Paul Martin and Richard Klein, eds., *Quaternary Extinctions.* Tucson: University of Arizona Press, 1984.

Pages 238–243. Madagascar and New Zealand extinction information comes from articles by R. Dewar and R. Cassels in Martin and Klein, eds., *Quaternary Extinctions.*

Chapter Eleven
Numbers

Pages 246–251. A summary of diversity and extinction rates can be found in David Raup, *Extinction, Bad Genes or Bad Luck?*

Page 251. *The New York Times* article appeared on August 20, 1991.

Pages 252–255. Much of the information about species diversity and extinction comes from E. O. Wilson, ed., *Biodiversity.* National Academy Press, 1988. This volume provides the best information about many of the issues discussed in Chapter 11. Chapters by Wilson, Ehrlich, Myers, Raup, Lugo, Raven, Erwin, Janzen, Taylor, Mittermeier, and Ramos were all used as information sources.

Pages 255–259. Additional information about tropical forests came from T. Whitmore, *An Introduction to Tropical Rain Forests.* Oxford University Press, 1990.

Pages 259–260. Discussion of human population size was abstracted from Strobe Talbott, *Time* magazine, December 16, 1991.

Page 260. The discussion of the NPP comes from P. Ehrlich, in Wilson, ed., op. cit.

Pages 261–269. Information about greenhouse warming and ozone depletions comes from C. Silver and R. DeFries, *One Earth, One Future,* National Academy Press, 1990, and from John Gribbin, *Hothouse Earth.* Grove Weidenfeld, 1990.

Chapter Twelve
Hope

Pages 276–278. Information about Costa Rica came from Dr. Gordon Orians.

Pages 278–279. E. O. Wilson, *The Diversity of Life.* Harvard University Press, 1992.

INDEX

Page numbers of illustrations and charts appear in italics.

ABOUT THE AUTHOR

Peter Douglas Ward is Professor of geological sciences, Adjunct Professor of zoology, and curator of paleontology at the University of Washington in Seattle. Dr. Ward has written four books, including *Beneath Puget Sound, In Search of Nautilus,* and *On Methuselah's Trail: Living Fossils and the Great Extinctions,* a Book-of-the-Month Club selection. Dr. Ward has also published numerous popular articles, including works in *Scientific American* and *Natural History,* and has been involved with many print, radio, and television features.